Applied Condition Monitoring 21

The book series Applied Condition Monitoring publishes the latest research and developments in the field of condition monitoring, with a special focus on industrial applications. It covers both theoretical and experimental approaches, as well as a range of monitoring conditioning techniques and new trends and challenges in the field. Topics of interest include, but are not limited to: vibration measurement and analysis; infrared thermography; oil analysis and tribology; acoustic emissions and ultrasonics; and motor current analysis. Books published in the series deal with root cause analysis, failure and degradation scenarios, proactive and predictive techniques, and many other aspects related to condition monitoring. Applications concern different industrial sectors: automotive engineering, power engineering, civil engineering, geoengineering, bioengineering, etc. The series publishes monographs, edited books, and selected conference proceedings, as well as textbooks for advanced students.

** Indexing: Indexed by SCOPUS, WTI Frankfurt eG, SCImago

Andrzej Puchalski · Bogusław Edward Łazarz ·
Fakher Chaari · Iwona Komorska ·
Radoslaw Zimroz
Editors

Advances in Technical Diagnostics II

Proceedings of the 7th International Congress on Technical Diagnostics, ICTD 2022, 14–16 September 2022, Radom, Poland

Springer

Editors
Andrzej Puchalski
Faculty of Mechanical Engineering
Kazimierz Pułaski University of Technology and Humanities
Radom, Poland

Fakher Chaari
National School of Engineers of Sfax
Sfax, Tunisia

Radoslaw Zimroz
Faculty of Geoengineering, Mining and Geology
Wrocław University of Science and Technology
Wroclaw, Poland

Bogusław Edward Łazarz
Faculty of Transport and Aviation Engineering
Silesian University of Technology
Gliwice, Poland

Iwona Komorska
Faculty of Mechanical Engineering
Kazimierz Pułaski University of Technology and Humanities
Radom, Poland

ISSN 2363-698X ISSN 2363-6998 (electronic)
Applied Condition Monitoring
ISBN 978-3-031-31718-7 ISBN 978-3-031-31719-4 (eBook)
https://doi.org/10.1007/978-3-031-31719-4

This Springer imprint is published by the registered company Springer Nature Switzerland AG
The registered company address is: Gewerbestrasse 11, 6330 Cham, Switzerland

Preface

The 7th International Congress on Technical Diagnostics (7th ICTD) was held at the University of Technology and Humanities in Radom on September 14–16, 2022.

ICTD is an international refereed Congress organized cyclically in Poland every four years dedicated to the advancement of the theory and practices in technical diagnostics. The idea of the Congress is to provide an international platform for all the world's industrial partners and researchers to share ideas and the latest concepts on the diagnostics of machines, vehicles, constructions, materials and industrial processes and maintenance of all kinds of technical facilities. 7th ICTD was an opportunity to highlight the digital transformation in the Industry 4.0 age.

The main speakers were outstanding experts in the field of diagnostics, both academic and industrial, namely

- Prof. Tomasz Krysiński, Airbus Helicopters
- Prof. Wiesław Jerzy Staszewski, AGH University of Science and Technology in Krakow
- Dr. Artur Pollak, APA Group
- Prof. Wojciech Batko, AGH University of Science and Technology in Krakow
- Prof. Zbigniew Dąbrowski, Warsaw University of Technology
- Prof. Jan Kiciński, Institute of Fluid-Flow Machinery Polish Academy of Sciences
- Prof. Stanisław Radkowski, Warsaw University of Technology
- Prof. Tadeusz Uhl, AGH University of Science and Technology, Space Technology Center.

The co-organizers of the Congress were Kazimierz Pulaski University of Technology and Humanities (Faculty of Mechanical Engineering), Polish Society of Technical Diagnostics and Polish Academy of Sciences: Committee on Acoustics—Section of Vibroacoustics and Committee on Machine Building, Future Industry Platform Foundation.

Participants submitted 84 papers. At least two referees reviewed all the chapters included in this book. We want to express our gratitude to all reviewers. Based on their opinions, 22 papers have been selected to be published as ICTD proceedings in the book series Applied Condition Monitoring.

We want to thank all the contributors for their work and for sharing the valuable research papers that have contributed to this volume.

We would also like to thank the Springer editorial and production team for their patience and valuable advice and support.

Andrzej Puchalski
Bogusław Łazarz
Iwona Komorska
Fakher Chaari
Radoslaw Zimroz

Organization

7th International Congress on Technical Diagnostics
14–16 September 2022, Radom – Poland
Webpage: https://ictd2022.uniwersytetradom.pl/

Organizing Committee

Andrzej Puchalski (Chair)
Iwona Komorska (Co-chair)
Marcin Wikło (Secretary)
Bartłomiej Drabik
Krzysztof Kołodziejczyk
Wojciech Kucharczyk
Marcin Migus
Przemysław Motyl
Krzysztof Olejarczyk
Małgorzata Wojtyniak

Scientific Committee

Wojciech Batko (Honorary Chair), Poland
Czesław Cempel (Honorary Chair), Poland
Zbigniew Dąbrowski (Honorary Chair), Poland
Bogusław Łazarz (Honorary Chair), Poland
Jérôme Antoni, France
Jose Antunes, Portugal
Andrew Ball, UK
Roman Barczewski, Poland
Tomasz Barszcz, Poland
Anna Borucka, Poland
Rafał Burdzik, Poland
Fakher Chaari, Tunisia
Xavier Chiementin, France
Wojciech Cholewa, Poland
Fulei Chu, China
Marco Cocconcelli, Italy
Robert Čep, Czech Republic
Mohamed El Badaoui, France
Spilios Fassois, Greece

Alfonso Fernández del Rincón, Spain
Marek Fidali, Poland
Piotr Folęga, Poland
Michael I. Friswell, UK
Krzysztof Górski, Poland
Andrzej Grządziela, Poland
Stephan Heyns, South Africa
Roman Kaczyński, Poland
Jan Kiciński, Poland
Grzegorz Klekot, Poland
Iwona Komorska, Poland
Łukasz Konieczny, Poland
Józef Korbicz, Poland
Theodoros Kosmanis, Greece
Zdzisław Kowalczuk, Poland
Aneta Krzyżak, Poland
Antoni Ligęza, Poland
Dragutin Lisjak, Croatia
Rafał Longwic, Poland
Jędrzej Mączak, Poland
Krzysztof Mendrok, Poland
Wojciech Moczulski, Poland
Andrzej Niewczas, Poland
Tomasz Nowakowski, Poland
Michał Pająk, Poland
Andrzej Pieczyński, Poland
Stanisław Radkowski, Poland
Minvydas Ragulskis, Lithuania
Riccardo Rubini, Italy
Adolfo Senatore, Italy
Jerzy A. Sładek, Poland
Aleksander Sładkowski, Poland
Wiesław J. Staszewski, Poland
Tomasz Stręk, Poland
Maciej Tabaszewski, Poland
Ryszard Tadeusiewicz, Poland
Mohamed Taktak, Tunisia
Anna Timofiejczuk, Poland
Tadeusz Uhl, Poland
Fernando Viadero-Rueda, Spain
Sławomir Wierzbicki, Poland
Adam Wiśniewski, Poland
Agnieszka Wyłomańska, Poland
Radosław Zimroz, Poland
Wojciech Żurowski, Poland

Congress Topics

7th ICTD 2022 concerns the theory and practices of technical diagnostics in:

- mechanical engineering,
- production engineering,
- automotive engineering,
- civil engineering and transport,
- computer engineering,
- materials engineering
- mining and energy,
- electronics and robotics,
- maritime and military industry,
- food industry,
- medicine

Contents

Ship Diesel Engine Fault Diagnosis Using Data Science and SVM Classifier ... 1
Michał Pająk, Marcin Kluczyk, Łukasz Muślewski, Dragutin Lisjak, and Davor Kolar

Laser Spot Thermography for Defect Detection in Composite Structure 13
Adam Machynia, Michał Sobczak, Ziemowit Dworakowski, and Jakub Roemer

Monitoring the Technical Condition of the Marine Turbine Engines by Analysis of Rundown Resonance Parameters 21
Andrzej Grządziela, Marcin Kluczyk, and Pero Vidan

Theoretical-Experimental Determination of CVT Power Losses Due to Rubber V-belt Vibrations 32
Waldemar Łatas and Adam Kot

Dynamic Model for Simulation of Phenomena Occurring in a Planetary Gear Operating in a Power Transmission System 42
Grzegorz Peruń

Acoustic Emission Monitoring for Wire Drawing Process 52
Enrique Caso, Pablo García-Fernández, Alfonso Fernández-del-Rincón, Ana De-Juan-De-Luna, Alberto Díez-Ibarbia, and Javier Sánchez-Espiga

Cycloidal Gearbox Model for Transient Analysis Implemented in Fortran with Constant Time Step 2nd Order Integrator 63
Roman Król

Experimental Setup for Nondestructive Testing of Composite Structures Using Laser Spot Thermography 75
Michał Sobczak, Adam Machynia, Ziemowit Dworakowski, and Jakub Roemer

Neural Network-Based Road Surface Condition Monitor Concept 85
Krzysztof Dobaj

Method of Fault Detection of Drive Systems Using Features of Park's Vector Hodograph 95
Krzysztof Roczek and Marek Fidali

Switched Reluctance Motor Modelling Methods 105
Jakub Lorencki and Stanisław Radkowski

Diagnostics of Logistics Processes Using Pallet 4.0® 115
Tomasz Korbiel

SYPIN: A System for Data Processing and Interpretation for Structural Health Monitoring 123
Ziemowit Dworakowski

Thermal Signature of Weld Defects Inside High Strength Steel Butt-Welded Joint Pipes 131
Massimiliano Pedot and Anna Timofiejczuk

Detection of Cyber Attacks in Electric Vehicles Using a Deep Neural Network 144
Marcin Gajdzik, Anna Timofiejczuk, Anna Gnacy-Gajdzik, and Piotr Przystałka

The Diagnostic Procedure of Internal Characteristics Identification of the Semi-active Automotive Shock Absorber Using a Numerical Model 154
Szymon Żymełka

Fault Diagnosis Using Artificial Neural Networks Trained Only on Signals from an Undamaged Machine 166
Paweł Pawlik, Konrad Kania, and Bartosz Przysucha

Algorithm of Sensory Network Routing Operating in Explosion Hazard Zones 176
Jerzy Jagoda, Krzysztof Stankiewicz, Joanna Rogala-Rojek, and Mariusz Hetmanczyk

Diagnostics of Worst-Case Execution Time Calculations for Automotive Embedded Systems 186
Kamil Sternal

Application of Torque Signal Analysis of Servomotors to Assess of Support System Condition of Industrial Machining Centre 194
Damian Augustyn and Marek Fidali

Fuzzy Condition Indicator of Spindle of Machining Center Based on Vibration Parameters 203
Damian Augustyn and Marek Fidali

Ship Diesel Engine Fault Diagnosis Using Data Science and SVM Classifier

Michał Pająk[1](✉), Marcin Kluczyk[2], Łukasz Muślewski[3], Dragutin Lisjak[4], and Davor Kolar[4]

[1] University of Technology and Humanities, Ul. Stasieckiego 54, 26-600 Radom, Poland
m.pajak@uthrad.pl
[2] Polish Naval Academy, Ul. Śmidowicza 69, 81-127 Gdynia, Poland
m.kluczyk@amw.gdynia.pl
[3] University of Science and Technology, Al. Kaliskiego 7, 85-796 Bydgoszcz, Poland
lukasz.muslewski@pbs.edu.pl
[4] University of Zagreb, Ivana Lučića 5, 10002 Zagreb, Croatia
{dragutin.lisjak,dkolar}@fsb.hr

Abstract. Main engine is the most significant element of reliability structure of the motor vessel. Its failure excludes the ship from the operation and leads to the most dangerous situation - loss of manoeuvrability. Therefore, accurate fault diagnosis should be key element of the modern maintenance strategies. In mechanical engineering the vibrations and acoustic signals recorded during the object operation is the most meaningful data used to identify the state of reliability. The diagnostics of damage to a marine diesel engine presented in the article is therefore carried out based on vibration and noise signals collected on each of the engine cylinders. It is proposed to identify the engine reliability state and the type of failure state using supervised machine learning. At first, the vibrations and noise signals were recorded during normal operation and during operation in five fault engine conditions. The recordings were then divided into parts by masking with a time window of calculated length. To model real situations where the number of available input data is limited, the number of signals were increased using shuffling and overlapping of masking windows. For each of received signal the characteristic values were calculated. Among the calculated values, the most significant features were selected. The obtained set of labelled samples was divided into learning and testing subsets. A subset of the learning data was used to identify the state of reliability and type of engine failure state, thanks to the use of machine learning. Eventually, the quality of the proposed identification method was checked using a subset of testing data.

Keywords: Fault Diagnosis · Data Science · Machine Learning · Ship Diesel Engine

1 Introduction

Motor vessel is a complex technical system, which, like any system, consists of many subsystems, but the most important from the point of view of safety is the propulsion system, because any failure of it may lead to a loss of manoeuvring speed. Moreover,

A. Puchalski et al. (Eds.): ICTD 2022, ACM 21, pp. 1–12, 2023.
https://doi.org/10.1007/978-3-031-31719-4_1

analysing the jurisprudence of maritime courts, it can be noticed that one of the main reasons for the loss of a vessels is the loss of its manoeuvring speed.

The crucial element of the reliability structure of the vessel propulsion system is the main engine. Thus, its technical condition should be subject to the continuous diagnostics.

One of the most meaningful diagnostic signals of rotating mechanical devices are vibration and noise signals. Therefore, in the presented work, the task of identifying the state of reliability of the research object was performed analysing the vibration and noise signals recorded during its operation.

The analysis of the vibrations and noise signals could be performed in frequency or time domain. Time domain analysis is used when it is difficult to obtain the value of rotating frequency or the values of operational parameters of mechanical parts. In such cases, some time domain features can be lost [1, 2]. In time domain, the classification-based fault diagnosis is used as a vibration analysis method. Among the most popular, linear discriminant analysis (LDA) [3], artificial neural network (ANN) [4], convolution neural network (CNN) [5], support vector machine (SVM) [6] or sparse representation based classification (SRC) [7] can be enumerated. In [8], a method for detecting bearing fault was proposed, based on the Resnet classifier with model data augmentation. The proposed method was tested on the real data and the trained Resnet classifier was able to identify the bearing conditions with 100% accuracy. In a study, Bhadane [9] used CNN for classification purposes feeding the CNN with the statistical features extracted from the vibration data. He found that CNN performed well in classifying bearing faults. SVM classifier was used to diagnose rolling bearing failures in [10] by hybrid approach where optimised SVM was combined with shift invariant sparse feature method. As a result of the studies, the accuracy of fault identification ranged from 93% to 96,2%, depending on the procedure used to optimise the classifier. In [11] on the basis of SRC, a method for classifying machine vibration signals was proposed, named Transform Domain Sparse Representation-based Classification (TDSRC). In TDSRC, the dictionary for sparse representation is not created with the raw samples, but with the transformation coefficients of the raw samples.

The diagnostics of faults in a marine diesel engine has been classified by the authors as an issue in which analysis in the time domain will be applied [12]. In the presented research, the machine-learning approach was used.

Vibration and noise signals were recorded on the object in a known reliability state, therefore the most appropriate way to build the identification model is supervised learning approach. Taking into consideration industrial applicability and simplicity of the machine learning model, SVM method was chosen for use.

At the initial stage of work, the procedure of the research performed to determine the reliability state and type of failure of the marine engine was established. It included operational tests, signal processing, machine-learning model formulation and optimization, and identification of reliability states and types of the inability states. It should be stressed that the procedure is a systemic approach to the considered issue and can be implemented to identify the reliability state of any rotating system.

According to the formulated procedure, the remaining part of the paper is organized. In Sect. 2 object of the research was characterized and the operational tests curried out to register vibration and noise signals of the selected engine were described. Next, the signals were analysed using Data Science techniques, as presented in Sect. 3. Then, the selected and transformed signals were used to build Machine Learning model and identify the reliability state and type of the failure of the tested marine engine. The description of this step can be found in Sects. 4 and 5. At the end of the paper, in Sect. 6, summary, conclusions, and planned future research were presented.

2 Research Object and Performed Operational Tests

The research object was marine engine Sulzer 6AL20/24 type produced by Zaklady Przemyslu Metalowego (ZPM) H.Cegielski in Poznan Poland. In Poland this type of engines are often used as a main propulsion on civilian ships and navy vessels (Fig. 1).

The Sulzer 6AL20/24 is a mid-speed, six-cylinder in-line, four-stroke, turbocharged marine engine. It is water-cooled, non-reversible engine with direct fuel injection. The ignition order is 1-4-2-6-3-5.

Fig. 1. Rescue ship ORP Zbyszko with Sulzer 6 AL 20/24 as the main propulsion

Operational tests were carried out on a laboratory stand, which is located in the Ship Engine Room Laboratory of the Mechanical and Electrical Engineering Faculty of the Polish Naval Academy in Gdynia. The tests consist in recording vibration and noise signals generated by the engine operating in the ability state and various inability states. The tests were performed with the same operating parameters values. The rotational speed was about 750 rpm and torque was 3.3 kNm. Vibration signals (vertical and horizontal direction) were measured using heads placed on each cylinder. The measurements of noise intensity were carried out using a microphone placed 200 cm from the engine.

The sampling frequency was 32768 Hz. During the tests, five different failures were simulated according to the following list:

- VC02 - Operation of an engine with input and output valves clearance reduced to 0.2 mm for cylinder 4 and output valves clearance increased to 1 mm for cylinder 1
- VC08 - Operation of an engine with increased input and output valves clearances to 0.8 mm for cylinder 4 and output valves clearance increased to 1 mm for cylinder 1
- CLE - Operation of an engine with output valves clearance increased to 1 mm for cylinder 1
- PUMP - Operation of an engine with simulated cylinder 1 injection pump failure - maximum combustion pressure dropped by about 20 and output valves clearances increased to 1 mm for cylinder 1
- INJ - Operation of an engine with injector of cylinder 1 opening pressure reduced to 23 MPa and output valves clearances increased to 1 mm for cylinder 1

For each reliability state, for each cylinder, vertical and horizontal vibrations and noise intensity were recorded. It gives 18 signals for each reliability state.

3 Data Analysis

The values of all 18 measured signals recorded for a specified time t formulate a measurement vector. Each recorded time history of the measured signals (18 signals in total) was divided into sets of 16384 consecutive measurement vectors. This resulted in 28 sets for states N (operation in ability state), VC02 and VC08, 29 sets for state CLE, 26 sets for state INJ and 13 sets for PUMP state. Form a machine-learning perspective, each set of measurement vectors is a labelled sample.

Based on literature studies [13–15] and own experience, it was found that the number of samples is not sufficient to prepare a machine-learning model. This situation is very common in real industrial systems where the amount of data is limited. Therefore, it was decided not to increase the length of the recorded time histories, but to solve the problem by augmenting the data using slicing and permutation in the form of random shuffling [16].

Random shuffling was used initially for each recorded time history. It consists in random rearranging of measurement vectors in a time series. As a result, new time histories were obtained. It multiplied the amount of data by two. Subsequently, slicing by time window with overlapping was executed. The length of the window was 16384 measurement vectors (assumed sample length) and the window shift was 10% of the window length (1638 measurement vectors). Each time history was extended by 16384 measurement vectors copied from its beginning to include all data in the resulting sets of measurement vectors. Thanks to the performed data augmentation, the number of measurement vector sets increased twenty times, which gives 562 sets for states N, VC02 and VC08, 582 sets for state CLE, 522 sets for state INJ and 262 sets for PUMP state.

After data augmentation, the characteristics of each signal were calculated according the following list [17]:

1. integral of signal,

2. mean value of signal,
3. energy of signal,
4. medium power of signal,
5. simple moment of the first order,
6. simple moment of the second order,
7. central moment of the first order,
8. central moment of the second order,
9. normalized simple moment of first order,
10. normalized simple moment of second order,
11. normalized central moment of first order,
12. normalized central moment of second order,
13. abscissa of the signal square gravity centre,
14. variance of the signal square,
15. equivalent diameter of signal,
16. mean width of signal.

Then, each set of the measurement vectors was replaced with the values of the calculated characteristics. In this way labelled data samples were obtained in the form of 288 features (16 characteristics multiplied by 6 cylinders multiplied by 3 signals: horizontal vibrations, vertical vibrations, and intensity of noise) and label describing the reliability state and the type of the failure for which the sample was recorded (1).

$$DS_i = \begin{bmatrix} V1:1,\ldots,V1:16, H1:1,\ldots,H1:16, N1:1,\ldots,N1:16, \\ \ldots, \\ V6:1,\ldots,V6:16, H6:1,\ldots,H6:16, N6:1,\ldots,N6:16 \\ RS \end{bmatrix} \tag{1}$$

where: DS_i – the i-th data sample, $Vj{:}k$ – the value of k-th characteristic calculated for j-th cylinder for vertical vibrations signal, $Hj{:}k$ – the value of k-th characteristic calculated for j-th cylinder for horizontal vibrations signal, $Nj{:}k$ – the value of k-th characteristic calculated for j-th cylinder for noise signal, RS – the reliability state.

The data samples were divided into two sets: the learning and the testing set with a ratio of 80/20. The samples for the testing set were randomly selected, however, the proportions of individual categories (reliability states) were maintained as in the total set.

In order to check the influence of data augmentation on the concentration of the characteristics, the concentration of each characteristic for each reliability state was calculated separately for original, transformed, and original+transformed samples using learning data set. The characteristic was treated as concentrated one if standard deviation (σ) of the characteristic value (D) in a group of samples (DSG) was less than 15% of the mean value (2).

$$\sigma_{D(DSG)} \leq \overline{\mathrm{D}}(\mathrm{DSG}) \cdot 0.15 \tag{2}$$

where: $D(DSG)$ - the characteristic of the signal, DSG - the group of the of the data samples.

Despite slightly lower concentration of characteristics for the transformed samples for some signals (especially noise signals) in relation to original ones for original and transformed samples together, the concentration was even higher. Therefore, it was concluded that the proposed augmentation may be applied in the classification issue under consideration.

Additionally, each concentrated characteristic was analysed in terms of identifying reliability states and types of inability states. It was assumed that the characteristic could be used to uniquely identify the reliability state if it meets the condition described by formula (3) and can be used to uniquely identify the type of the inability state if it meets the condition described by the formula (4) for all inability states.

$$\sigma_{D(DSG_{IA})} + \sigma_{D(DSG_{A})} \leq \left|\overline{\mathrm{D}}(\mathrm{DSG}_{IA}) - \overline{\mathrm{D}}(\mathrm{DSG}_{A})\right| \quad (3)$$

where: *D(DSG)* - the characteristic of the vibration signal, DSG_{IA} - the group of the data samples for inability states, DSG_A - the group of the data samples for ability states.

$$\sigma_{D(DSG^{i}_{IA})} + \sigma_{D\left(DSG^{j}_{IA}\right)} \leq \left|\overline{\mathrm{D}}\left(DSG^{i}_{IA}\right) - \overline{\mathrm{D}}\left(DSG^{j}_{IA}\right)\right| \quad (4)$$

where: *D(DSG)* - the characteristic of the vibration signal, ${DSG_{IA}}^{i}$ - the group of the data samples for *i*-th inability states.

In case of identifying the states of reliability, there were two concentrated characteristics, which fulfil condition (3): energy and medium power of horizontal vibrations recorded on cylinder 4. In case of type of inability state identification there were no concentrated characteristic, which meet the condition (4) calculated between every types of inability states. It means that the considered classification issue is not a linear one thus any linear classifier cannot be used in the conducted research.

The next step of the data analysis was the reduction of the characteristics number. It was done by correlation analysis between every characteristic using the Pearson correlation coefficient [18] calculated for learning data set (5).

$$r_{ij} = \frac{\sum_{k=1}^{card(DS_{learning})} \left(D_i^k - \overline{D}_i\right) \cdot \left(D_j^k - \overline{D}_j\right)}{\sqrt{\sum_{k=1}^{card(DS_{learning})} \left(D_i^k - \overline{D}_i\right)^2} \cdot \sqrt{\sum_{k=1}^{card(DS_{total})} \left(D_j^k - \overline{D}_j\right)^2}} \quad (5)$$

where: r_{ij} – the sample Pearson correlation coefficient between *i*-th and *j*-th characteristic, $DS_{learning}$ – the learning set of the data samples, ${D^K}_i$ – the value of the *i*-th characteristic for *k*-th data sample.

In the figure (Fig. 2) the diagrams of exemplary correlated and not correlated characteristics can be found. If the value of the correlation coefficient between two characteristics was greater or equal to 0.8 one of them was excluded from further considerations. In this way, the number of the characteristics was reduced to 157 according to the following list:

- For vertical vibrations, characteristics number 1, 3, 7, 8, 9, 11, 12, 13, 14, 15 for each cylinder
- For horizontal vibrations, characteristics number 1, 3, 7, 8, 9, 11, 12, 13, 14, 15 for each cylinder

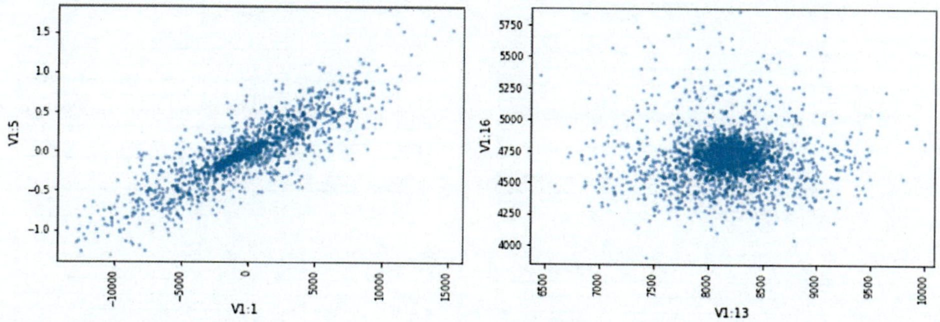

Fig. 2. V1:5 = f(V1:1) - the exemplary strongly correlated characteristics ($r_{V1:5,V1:1} = 0.85$), V1:16 = f(V1:13) - exemplary weakly correlated characteristics ($r_{V1:16,V1:13} = 0.026$)

- For noise signal, characteristics number 9, 11, 12, 13, 14, 15 for each cylinder and characteristic number 1 for cylinder 1

Since most machine-learning algorithms deal poorly with numerical attributes in various ranges of the scale, the last step of the data preparation was the standardization of samples. It was accomplished by mean value subtraction and division by the standard deviation according to formula (6)

$$D_i'^k = \frac{D_i^k - \overline{D}_i}{\sigma_{D_i}} \tag{6}$$

where: D'^K_i – the value of the i-th characteristic for k-th data sample after standardization, D_i – the i-th characteristic of the signal, D^K_i – the value of the i-th characteristic for k-th data sample, σ_{Di} – the standard deviation of the i-th characteristic of the signal.

Then, the data samples prepared in the way presented above were used to identify the reliability state and the type of inability state using machine-learning modelling.

4 Reliability and Type of Inability State Identification

The main purpose of the performed research is to identify the reliability state of the considered engine by analysing vibration and noise signals. As already mentioned, the analysis was performed in time-domain. In the Fig. 3 the exemplary vibrations time histories for ability and all simulated inability states are presented. Looking at the figure, it can be noticed that vibrations time histories are slightly different, however it is not obvious to distinguish the reliability state on this basis. Therefore, it was decided to use one of the machine learning classifiers to accomplish this task.

The support vector machine (SVM) is a powerful and versatile machine-learning model capable of performing linear or non-linear classification. It is one of the most popular machine-learning solutions commonly used in engineering applications [19–21]. It is especially useful for classifying complex, small or medium-sized data sets.

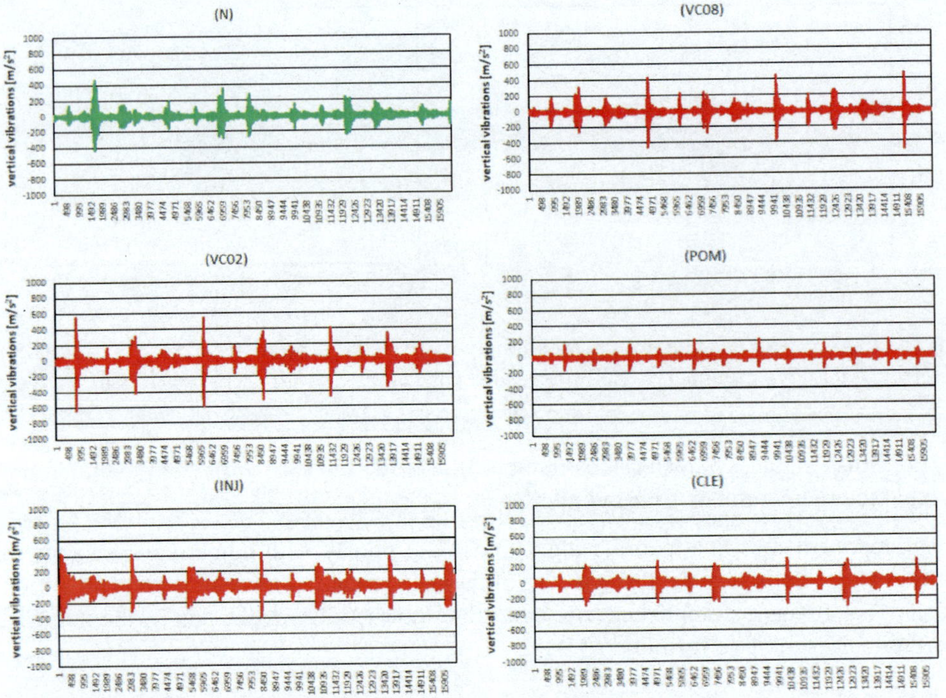

Fig. 3. Exemplary vibrations time histories

As described above the reliability state and type of inability state identification is a complex classification problem where the cardinality of the data samples set is small. Therefore, it was decided to use SVM modelling to solve the considered problem.

The fundamental idea behind SVM modelling is to find the optimal hyperplane to separate data samples of different classes. During the analysis of the data (described in previous chapter) it was found that the reliability state and the type of the inability state identification is an issue for which a linear classification cannot be used. Therefore, non-linear variant of SVM should be applied. This was achieved by polynomial kernel implementation.

The SVM classifier is the binary one, therefore, to identify several classes of data samples (one ability state and five types of inability state) the one-versus-one (OvO) strategy was used. It consists in learning $N(N-1)/2$ classifiers to compare classes one to each other, where N is the number of classes. The obtained class of the analysed sample is the one that wins in most comparisons.

The complete procedure of the ability state and the type of inability states identification is illustrated in the diagram shown in Fig. 4.

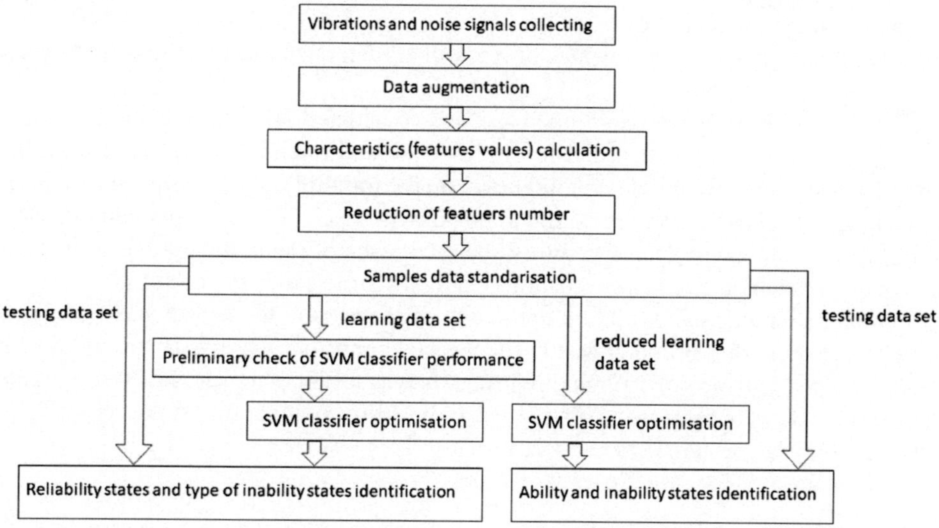

Fig. 4. Procedure of the conducted research

5 Results

In order to preliminary check the performance of non-linear SVM classifier with polynomial kernel (P-SVM), the k-fold cross-validation procedure was used. According to the procedure, the learning set is randomly divided into *k* folds, then the model is trained and evaluated *k* times (each time a different subset is selected to evaluate the performance of the model, and the remaining *k-1* are used for training). Finally, *k* evaluations results are obtained. In the study, *k* was equal to 3 and the performance was assessed on accuracy (correct classifications to all classifications ratio). For default values of the model hyper-parameters the mean accuracy was 83%.

P-SVM model has several hyper-parameters where the most significant are the degree of kernel, the constant term of the kernel and the regularization parameter, which is used to avoid model overfitting. The strength of the regularization is inversely proportional to this parameter. In order to determine the best values of hyper-parameters grid search method was used. Thanks to it, for all combinations of entered values of hyper-parameters the model performance was evaluated by k-fold cross-validation procedure. In the research, the following values of the hype-parameters were used:

- the degree of kernel: 3, 6, 9
- the constant term of the kernel: 1, 10, 100
- the regularization parameter: 1, 3, 5

Given the number of folds equal to 3 the best combination was obtained in form: 3, 1, 1 respectively. For the found values of the hyper-parameters, the mean accuracy of the model obtained by k-fold cross-validation procedure was 0.999. Therefore, it was decided that the model is well trained, and its performance can be checked using the testing data set. The obtained model accuracy over testing data set was 100%. Typically, several measures are used to analyse classification quality, such as confusion matrix, precision, selectivity, recall, or F1 score. However, in case when accuracy is 100% the confusion matrix has non-zero values only on main diagonal and remaining measures have theirs highest values.

In the last step of the research, the case was considered when in the learning data set there are only samples recorded for ability state and one inability state and in the testing data set there are samples for unknown on the training stage inability states. Five reduced training datasets were prepared to begin with. Each of them contained samples for ability state and one selected inability state. For each of them, the optimal values of the model hyper-parameters were determined using grid search method. In each case, the optimal kernel degree was 3, the optimal constant term of the kernel was 1 and the optimal regularization parameter was 1. Then, each learning set was used to train P-SVM model. The performance of the resulting models was tested using the full testing data set, where labels were changed to distinguish only ability and inability state. As a result, 100% accuracy was obtained in each case.

6 Summary and Conclusions

In the presented studies, a new universal procedure for identifying reliability states was proposed. The procedure is a systemic approach to the issue of the ability and the type of the inability states identification of a complex technical system.

The proposed method was used to identify the ability state and the type of the inability state of marine engine of Sulzer 6AL20/24 type. Using data science techniques to analyse vibrations and noise signals recorded during operational tests and machine-learning modelling in form of non-linear SVM classifier with polynomial kernel, 100% accuracy of the identification process was obtained. Moreover, this method was used to distinguish between ability and inability states when in testing data set signals recorded for unknown inability state were present. In this case also 100% accuracy of the identification was achieved.

As a result of the research 100% accurate identification of the state of reliability was obtained. Thus, it has been proven that the implementation of the data science techniques and machine-learning modelling in the complex technical systems diagnostics brings high quality solutions.

The quality of the obtained solution encourages conducting further research in this area. It is planned to verify the effectiveness of the proposed approach in the case of vibrations and noise signals recorded during engine operation in the workplace. The second direction of further studies is to apply the method in the case when learning data set consists of signals recorded at the laboratory stand and testing data set consists of signals recorded at the engine workplace. Having achieved high accuracy in this configuration, the method will be ready for practical use. Finally, a combination of supervised and unsupervised learning methods implementation will be performed. The idea is to train a selected machine-learning model through fully labelled samples (known ability and inability states) and use it to identify known and unknown inability states by detecting new clusters of samples.

References

1. Tseng, M.L., Wu, K.J., Ma, L., Kuo, T.C., Sai, F.: A hierarchical framework for assessing corporate sustainability performance using a hybrid fuzzy synthetic method-DEMATEL. Technol. Forecast. Soc. Change. **144**, 524–533 (2019). https://doi.org/10.1016/j.techfore.2017.10.014
2. Xuan, Q., et al.: Automatic Pearl classification machine based on a multistream convolutional neural network. IEEE Trans. Ind. Electron. **65**, 6538–6547 (2018). https://doi.org/10.1109/TIE.2017.2784394
3. Mboo, C.P., Hameyer, K.: Fault diagnosis of bearing damage by means of the linear discriminant analysis of stator current features from the frequency selection. IEEE Trans. Ind. Appl. **52**, 3861–3868 (2016). https://doi.org/10.1109/TIA.2016.2581139
4. Hu, N., Chen, H., Cheng, Z., Zhang, L., Zhang, Y.: Fault diagnosis for planetary gearbox based on EMD and deep convolutional neural networks. Jixie Gongcheng Xuebao/J. Mech. Eng. **55**, 9–18 (2019). https://doi.org/10.3901/JME.2019.07.009
5. Kolar, D., Lisjak, D., Pająk, M., Gudlin, M.: Intelligent fault diagnosis of rotary machinery by convolutional neural network with automatic hyper-parameters tuning using Bayesian optimization. Sensors **21**, 2411 (2021). https://doi.org/10.3390/s21072411
6. Agrawal, P., Jayaswal, P.: Diagnosis and classifications of bearing faults using artificial neural network and support vector machine. J. Inst. Eng. (India): Ser. C **101**(1), 61–72 (2019). https://doi.org/10.1007/s40032-019-00519-9
7. Wang, H., Ren, B., Song, L., Cui, L.: A novel weighted sparse representation classification strategy based on dictionary learning for rotating machinery. IEEE Trans. Instrum. Meas. **69**, 712–720 (2020). https://doi.org/10.1109/TIM.2019.2906334
8. Qian, L., Pan, Q., Lv, Y., Zhao, X.: Fault detection of bearing by ResNet classifier with model-based data augmentation. Machines **10**, 521 (2022). https://doi.org/10.3390/machines10070521
9. Bhadane, M., Ramachandran, K.I.: Bearing fault identification and classification with convolutional neural network. In: International Conference on Circuit, Power and Computing Technologies (ICCPCT), pp. 1–5 (2017). https://doi.org/10.1109/ICCPCT.2017.8074401
10. Yuan, H., Wu, N., Chen, X., Wang, Y.: Fault diagnosis of rolling bearing based on shift invariant sparse feature and optimized support vector machine. Machines **9**, 98 (20212). https://doi.org/10.3390/machines9050098
11. Fajun, Y., Fuling, F., Shuanghong, W., Fengxing, Z.: Transform-domain sparse representation based classification for machinery vibration signals. J. Vibroengineering **20**, 979–987 (2018). https://doi.org/10.21595/jve.2017.18865
12. Pająk, M., Muślewski, Ł., Landowski, B., Kałaczyński, T., Kluczyk, M., Kolar, D.: Identification of reliability states of a ship engine of the type Sulzer 6AL20/24. SAE Int. J. Engines. **15**, 03-15-04-0028 (2021). https://doi.org/10.4271/03-15-04-0028
13. Vasegh, M., Sharifi Miavaghi, A.: A novel method for flexible solar energy generation system fault detection using optimally structured convolution neural networks. SSRN Electron. J. (2021). https://doi.org/10.2139/ssrn.3994263
14. Pan, J., Qu, L., Peng, K.: Sensor and actuator fault diagnosis for robot joint based on deep CNN. Entropy **23**, 751 (2021). https://doi.org/10.3390/e23060751
15. Nam, J., Kang, J.: Classification of chaotic squeak and rattle vibrations by CNN using recurrence pattern. Sensors **21**, 8054 (2021). https://doi.org/10.3390/s21238054
16. Steven Eyobu, O., Han, D.S.: Feature representation and data augmentation for human activity classification based on wearable IMU sensor data using a deep LSTM neural network. Sensors **18**, 2892 (2018). https://doi.org/10.3390/s18092892
17. Jerzy, S.: Fundamentals of the Signal Theory. WKŁ, Warszawa (2007)

18. Temizhan, E., Mirtagioglu, H., Mendes, M.: Which correlation coefficient should be used for investigating relations between quantitative variables? Am. Acad. Sci. Res. J. Eng. Technol. Sci. **85**, 265–277 (2022)
19. Nafees, A., et al.: Forecasting the mechanical properties of plastic concrete employing experimental data using machine learning algorithms: DT, MLPNN, SVM, and RF. Polymers (Basel) **14**, 1583 (2022). https://doi.org/10.3390/polym14081583
20. Wu, Y., Li, S.: Damage degree evaluation of masonry using optimized SVM-based acoustic emission monitoring and rate process theory. Measurement **190**, 110729 (2022). https://doi.org/10.1016/j.measurement.2022.110729
21. Wang, Z., He, X., Shen, H., Fan, S., Zeng, Y.: Multi-source information fusion to identify water supply pipe leakage based on SVM and VMD. Inf. Process. Manag. **59**, 102819 (2022). https://doi.org/10.1016/j.ipm.2021.102819

Laser Spot Thermography for Defect Detection in Composite Structure

Adam Machynia(✉), Michał Sobczak, Ziemowit Dworakowski, and Jakub Roemer

Department of Robotics and Mechatronics, AGH University of Science and Technology, Krakow, Poland
machynia@agh.edu.pl

Abstract. Thermography is a significant part of nondestructive testing and is favorably used in composite materials examination. Specifically, laser thermography, which is one of the active thermography methods, utilizes heat excitation achieved by means of a laser. This technique grants precise and repetitive measurements which contributes to effective data processing and interpretation. In this paper defect detection technique based on active laser thermography measurements is presented. The main part of this work is introducing the processing routine for thermograms' sequences which ultimately leads to fault detection. Automatic analysis of thermograms utilizing image processing methods allows for straightforward feature extraction for further examination. Having obtained meaningful and concise representations of measurements, machine learning techniques are used for classification purposes. A multi-layer perceptron, a classical neural network, and support vector machine are applied for this task. Finally, a specific area of the examined sample is indicated as faulty or flawless. The method is presented in the case of the composite plate containing prepared defects imitating delamination at various depths.

Keywords: laser spot thermography · defect detection · composite structures · machine learning

1 Introduction

There are many non-destructive testing (NDT) methods that nowadays are widely used and that are well standardized and developed. However, with increasing development in high-performance materials, new detection methods have to be developed in order to meet ever-increasing safety and reliability standards. Among many methods for NDT thermography offers a vast range of advantages for defect detection not only on large surfaces but also on smaller precise samples. It bases on the analysis of thermal heat patterns in the inspected materials. Thermography can be divided into two main branches. The first one is passive thermography, where no additional heat is provided to inspected object. A simple thermographic image is taken and analyzed. This method is widely used

A. Puchalski et al. (Eds.): ICTD 2022, ACM 21, pp. 13–20, 2023.
https://doi.org/10.1007/978-3-031-31719-4_2

for example in the detection of heat flows in buildings [1,2] or for detection of faulty photovoltaic cells [3]. Active thermography, on the other hand, requires some sort of excitation. It can utilize heat sources in form of optical excitation, mechanical or other forms such as Eddy currents [10]. Vibrothermography is a method that uses friction to generate heat based on provided mechanical excitation [9]. Inspections can be conducted in the form of a sequence recording for a single area [12], point-by-point scan [5] or using continuous scanning [7] method, where the camera is moving with respect to the sample.

Optical excitation can be generated using flash [14], halogen lamps[6] or laser beam [11]. Lamp excitation performs well during inspections of large surfaces, as they are fast. On the other hand, non-uniform heat distribution and reflections need to be taken into consideration. The laser gives the possibility to conduct NDT measurements for precise samples, with high resolution as it gives the possibility to precisely control parameters such as delivered energy and laser pulse length. The challenge for these methods is that due to more inspection points the data need to be composed and inspection takes longer. The great advantage is the possibility to detect cracks perpendicular to the surface and good resolution of inspection.

In this work, a novel defect detection method is presented. Utilizing laser spot thermography, geometric features of binarized thermograms are analyzed. Further, in this approach machine learning techniques such as multilayer perceptron (MLP) and support vector machine (SVM) are used for classification purposes.

2 Methodology

This section covers the methodology of measurements and data processing routine. The whole processing procedure is implemented in Python.

2.1 Experimental Setup

For the purposes of laser spot thermography, a special test rig was created. The system's main controller is working on a PXIe-8821 computer with NI Linux Real-Time operating system (RTOS). The usage of RTOS allows for precise synchronization of control signals. User-interface is located on a PC computer connected to the main controller. GUI is implemented using National Instruments Labview software. 120 W Limo laser source 120-F400-DL808-AV5-A is responsible for generating laser impulse. It is controlled by the CS412 Amtron laser controller, which gives the possibility to set, among others, laser power and pulse duration. At the end of the fiber-optics, laser head Limo PH-LD80-R-35-0 with collimating lenses creates a laser spot of diameter approx. 10 mm. The thermographic sequence is recorded using FLIR A655SC infrared camera. It is capable of recording thermal images with a resolution of 640×480 pixels and with 50 frames per second. Camera and laser optics are placed stationary on the platform in front of the moving sample. The precise position of the sample is ensured by two Standa 8MTL1401-300 Direct Drive Linear Translation Stages controlled by the SPiiPlusCMNT motor controller. The workspace of the test rig is 300×300 mm. Thanks to the synchronization of the control signals, and precise positioning of the camera with respect to the sample, results are repeatable allowing for better performance of defect detection algorithms

The examination procedure is tested and presented on the carbon fiber reinforced polymer (CFRP) sample, which contains four Teflon inserts imitating defects and placed at two different depths. The input for the defect detection procedure is a thermographic sequence containing 244 frames. Each file includes a measurement representing a specific area. The unit is set to Celsius degrees.

2.2 Image Sequence Processing

The procedure starts with independent processing of each file containing a thermographic sequence. Frames from 2nd to 5th after averaging forms background. Next, the video is restricted to a set of frames – from 20th to 100th, which represents the first part of cooling of the sample. From each frame of this subset, the background is subtracted in order to enhance laser spot visualization. Then, thermal images are binarized with a constant threshold t_b, experimentally set to three. Thus, a sequence of binary images is obtained. This part of processing is schematically shown in Fig. 1.

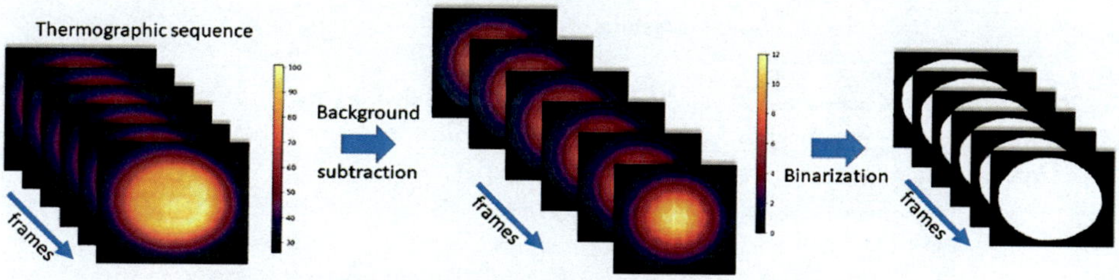

Fig. 1. Scheme of processing of thermographic sequence: background subtraction and binarization.

2.3 Features Extraction

All binary images are processed one by one. Geometric features of the region representing the laser spot are extracted. For this purpose scikit-image package for Python is utilized [13]. The extracted features are as follows.

1. Area – the number of pixels constituting the region.
2. Axis major length – as for the ellipse corresponding to the region in terms of normalized second central moments.
3. Axis minor length – as for the ellipse corresponding to the region in terms of normalized second central moments.

4. Eccentricity – as for the ellipse corresponding to the region in terms of normalized second central moments, the proportion of the distance between the focal points and the axis major length.
5. Maximal Feret diameter – the maximal distance between the two parallel planes containing the pixels located on the border of the region.
6. Perimeter – counted for border pixels of the region.

Extracted features are graphically presented in the Fig. 2.

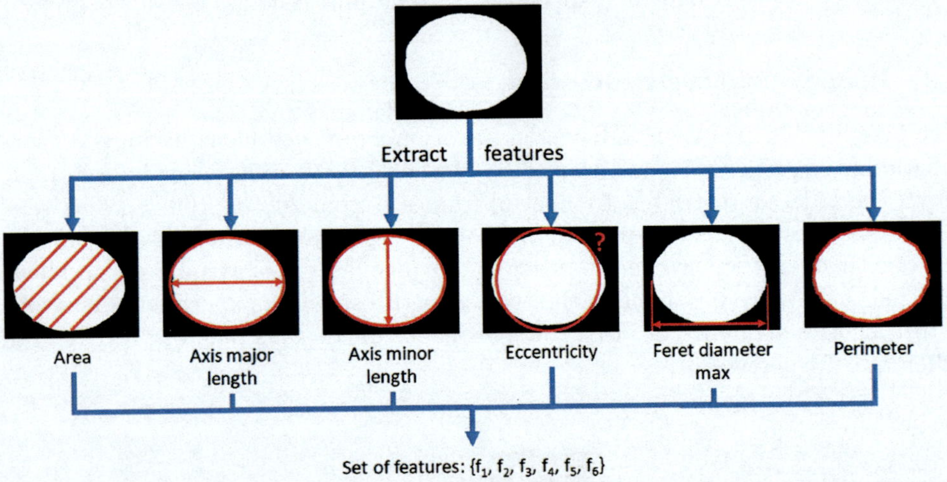

Fig. 2. Features vector consists of six parameters which are: area, axis major length, axis minor length, eccentricity, maximal Feret diameter, and perimeter. Features extraction is crucial step for further data processing.

Having obtained six geometric features of the laser spot's binary image for each frame and measuring point, they are organized into the features vector f of length equal to six. Thus, there are prepared vectors in total number equal to frames number n_f multiplied by measuring points number n_{mp}. These vectors are stacked together in order to prepare a concise representation of the whole processed data. Firstly, the vectors obtained for one measuring point, so the consecutive frames of a thermographic sequence, are joined lengthwise. This results in a vector of length $6 \cdot n_f$. Next, such vectors for each measuring point are considered as rows in a newly created matrix of size $n_{mp} \times 6 \cdot n_f$. Therefore, each row represents one measuring point and every six columns corresponds with a specific frame in a thermographic sequence. Schematically data stacking is presented in Fig. 3.

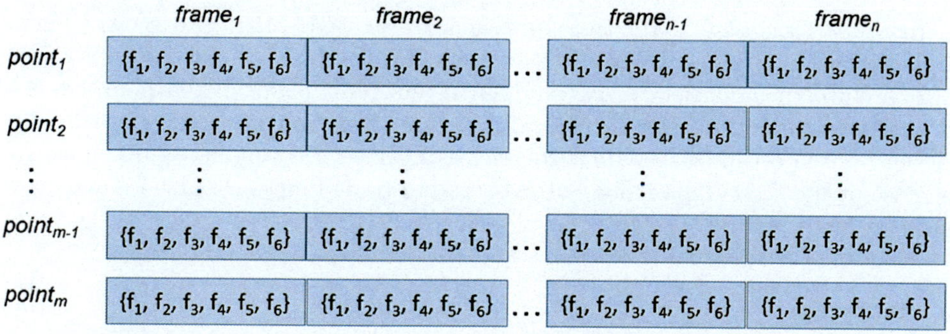

Fig. 3. The processed data matrix is composed of feature vectors for each frame and measuring points.

2.4 Classification

Each measuring point is labeled as an intact or damaged area. However, on the defects' borders, there are regions containing partially damaged and intact area, therefore they are excluded from the training and performance assessment procedure, and are treated as unlabeled. The ground truth map of a sample is shown in Fig. 4. The damaged state is represented by roughly only 4% of data, which results in a significant class imbalance.

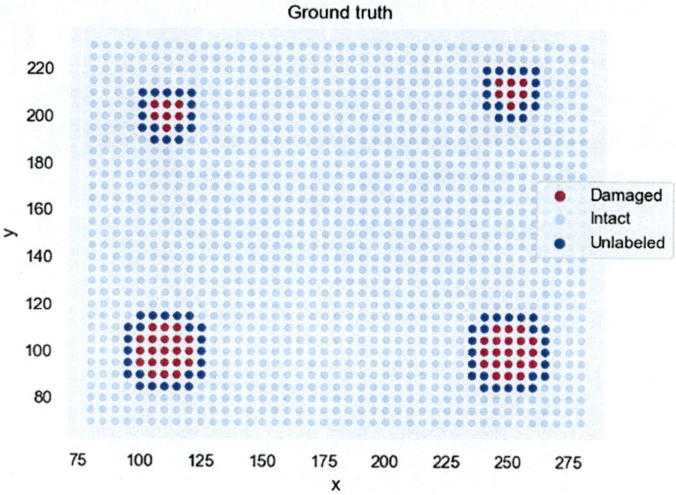

Fig. 4. The ground truth map of a sample.

In the first step, data is scaled into the [0, 1] range. Next, data is split into training, test, and validation sets in a ratio of 0.6, 0.25, and 0.15. In order to overcome the class imbalance issue, the random oversampling method is applied to the training data set [4].

In order to classify data two approaches are used: MLP and SVM. Their implementation utilizes *scikit-learn* package for Python [8]. In the first case, MLP architecture consists of two hidden layers built up from ten neurons. As an activation function, the rectified linear unit function is used. A stochastic gradient-based optimizer – *adam* solver – is utilized for weights adjustment. In the case of the second classifier – SVM – radial basis function (RBF) is selected as the kernel function.

3 Results

The results of classification using MLP are presented in Fig. 5(a). The light blue color indicates the prediction of an intact state and defect predictions are marked with magenta. In Fig. 5(b) the confusion matrix for test set is shown. The false positive rate is very low and the false negative rate remains satisfying. The unlabeled data needs to be excluded from the direct assessment, however, the predictions in these data points are the algorithm's asset.

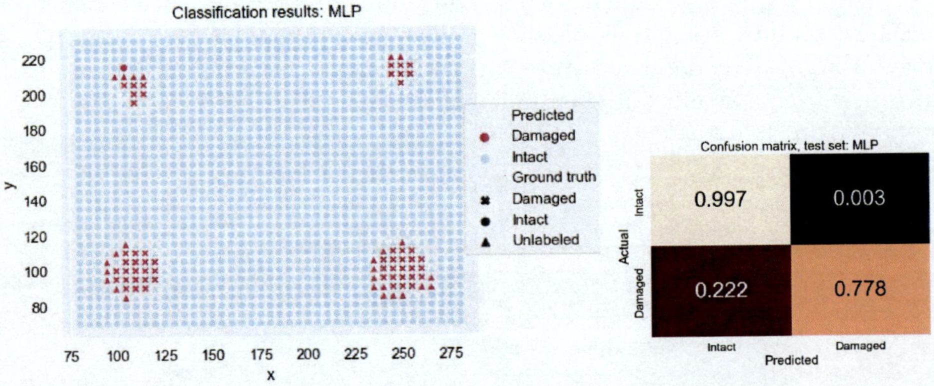

(a) (b)

Fig. 5. The results of classification with the use of MLP: (a) the predictions map compared to the ground truth, (b) the confusion matrix.

Similarly, the results obtained utilizing the SVM classifier are shown in Fig. 6(a) and the confusion matrix for test set is presented in Fig. 6(b). The false positive rate is at the exactly same very low level, while the false negative rate is even lower compared to the results obtained with MLP. Considering the whole sample, not only the test set, only two data points were misclassified as false positives. Generally, results obtained with both classifiers are quite similar.

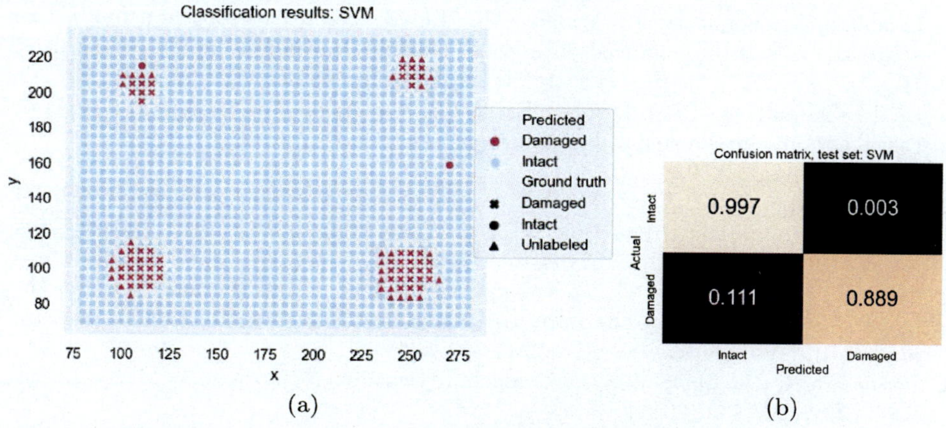

Fig. 6. The results of classification with the use of SVM: (a) the predictions map compared to the ground truth, (b) the confusion matrix.

4 Summary and Conclusions

The proposed processing routine is based on the analysis of binary images of thermograms, especially considering the geometry of heat transfer after laser spot excitation. Due to feature extraction, a proper machine learning application is possible. Assessment of the sample condition is performed utilizing MLP and SVM, which allows proper indication of faulty regions. As an extension of the proposed method, other features might be considered. Also, an inspection of different materials and defects might be conducted.

Acknowledgment. The research has been financed within the scope of project no. 0001/L-11/2019 "Laser thermography testing for damage detection in composite structures" financed by the National Center for Research and Development, Poland. Part of the work was also financed by the AGH University of Science and Technology, WIMiR, research grant no. 16.16.130.942/GD/2022.

References

1. Barreira, E., Almeida, R.M., Moreira, M.: An infrared thermography passive approach to assess the effect of leakage points in buildings. Energy Build. **140**, 224–235 (2017). https://doi.org/10.1016/j.enbuild.2017.02.009
2. Garrido, I., Lagüela, S., Otero, R., Arias, P.: Thermographic methodologies used in infrastructure inspection: a review-data acquisition procedures. Infrared Phys. Technol. **111**, 103481 (2020). https://doi.org/10.1016/j.infrared.2020.103481
3. Kandeal, A., Elkadeem, M., Kumar Thakur, A., Abdelaziz, G.B., Sathyamurthy, R., Kabeel, A., Yang, N., Sharshir, S.W.: Infrared thermography-based condition monitoring of solar photovoltaic systems: a mini review of recent advances. Solar Energy **223**, 33–43 (2021). https://doi.org/10.1016/j.solener.2021.05.032

4. Lemaître, G., Nogueira, F., Aridas, C.K.: Imbalanced-learn: a python toolbox to tackle the curse of imbalanced datasets in machine learning. J. Mach. Learn. Res. **18**(17), 1–5 (2017). http://jmlr.org/papers/v18/16-365.html
5. Li, T., Almond, D., Rees, D.A.S., Weekes, B.: Crack imaging by pulsed laser spot thermography. In: Journal of Physics: Conference Series, vol. 214, p. 012072 (2010). https://doi.org/10.1088/1742-6596/214/1/012072
6. Luo, Q., Woo, W.L., Yang, Y.: Temporal and spatial deep learning network for infrared thermal defect detection. NDT & E Int. **108**, 102164 (2019). https://doi.org/10.1016/j.ndteint.2019.102164
7. Montinaro, N., Cerniglia, D., Pitarresi, G.: Flying laser spot thermography for the inspection of aerospace grade fibre metal laminates, pp. 206–210 (2017). https://doi.org/10.1109/MetroAeroSpace.2017.7999565
8. Pedregosa, F., et al.: Scikit-learn: machine learning in python. J. Mach. Learn. Res. **12**, 2825–2830 (2011)
9. Pieczonka, L., Aymerich, F., Brozek, G., Szwedo, M., Staszewski, W., Uhl, T.: Modelling and numerical simulations of vibrothermography for impact damage detection in composites structures. Struct. Control Health Monit. **20**, 626–638 (2013). https://doi.org/10.1002/stc.1483
10. Rellinger, T., Underhill, P., Krause, T., Wowk, D.: Combining eddy current, thermography and laser scanning to characterize low-velocity impact damage in aerospace composite sandwich panels. NDT & E Int. **120**, 102421 (2021). https://doi.org/10.1016/j.ndteint.2021.102421
11. Rodríguez-Aseguinolaza, J., Colom, M., González, J., Mendioroz, A., Salazar, A.: Quantifying the width and angle of inclined cracks using laser-spot lock-in thermography. NDT & E Int. **122**, 102494 (2021). https://doi.org/10.1016/j.ndteint.2021.102494
12. Saeed, N., King, N., Said, Z., Omar, M.: Automatic defects detection in cfrp thermograms, using convolutional neural networks and transfer learning. Infrared Phys. Technol. **102**, 103048 (2019). https://doi.org/10.1016/j.infrared.2019.103048
13. van der Walt, S., et al.: The scikit-image contributors: scikit-image: image processing in Python. PeerJ **2**, e453 (2014). https://doi.org/10.7717/peerj.453
14. Zhou, J., Du, W., Yang, L., Deng, K., Addepalli, S., Zhao, Y.: Pattern recognition of barely visible impact damage in carbon composites using pulsed thermography. IEEE Trans. Ind. Inf. **18**(10), 7252–7261 (2022). https://doi.org/10.1109/TII.2021.3134184

Monitoring the Technical Condition of the Marine Turbine Engines by Analysis of Rundown Resonance Parameters

Andrzej Grządziela[1], Marcin Kluczyk[1(✉)], and Pero Vidan[2]

[1] Mechanical-Electrical Faculty, Polish Naval Academy, Śmidowicza 69 Street, 81127 Gdynia, Poland
{a.grzadzeila,m.kluczyk}@amw.gdynia.pl

[2] Faculty of Marine Studies, University of Split, Ruđera Boškovića 37, Street, 21000 Split, Croatia
pero.vidan@pfst.hr

Abstract. The mechanical complexity of marine turbine engines is very high, and they operate in challenging conditions. This determines the need to monitor their technical condition, especially since possible failures due to high rotational speeds of rotating elements have tragic consequences. Therefore, the authors propose a diagnostic method for marine turbine engines based on periodic measurements of their vibration parameters. The basis for the proper functioning of the proposed method is the possession by diagnosticians of a set of reference data relating to a fully operational engine, preferably a new specific copy. The resonance parameters recorded during the rundown of the rotor system were used, which eliminates the influence of the combustion process on the recorded vibration signals. The tests were carried out on four engines of the same type in continuous operation..

Keywords: marine turbine engine · diagnostics · resonance · technical condition

1 Introduction

Marine turbine engines are used due to their very high susceptibility to load changes. This is a very desirable feature in marine propulsion systems, where there are often rapid changes in the rotational speed of the main propulsion engines. However, this results in challenging operating conditions for these engines, which determines the need to use diagnostic systems that allow early detection of changes in technical conditions. These systems are commonly known as Health Monitoring Systems (HMS). Manufacturers of modern marine turbine engines offer, along with the engine its stationary online diagnostic system. Often, such a system, using satellite communication, sends diagnostic data to the manufacturer's server, where they are subjected to thorough expert analysis. Diagnostic data may contain a lot of important and sensitive information that unauthorized persons can use to obtain information about the strategy of using the warship on which they are mounted. Therefore, navies often use online systems with no data transfer or, especially in the case of older engines, offline systems. They consist in periodic

A. Puchalski et al. (Eds.): ICTD 2022, ACM 21, pp. 21–31, 2023.
https://doi.org/10.1007/978-3-031-31719-4_3

measurements of diagnostic parameters and drawing conclusions about the technical condition of the engine on this basis. This paper considers the use for the diagnostics of marine turbine engines the vibration parameters recorded especially during the rotor coast down from the moment of cutting off the fuel supply to the injectors. Such an approach makes it possible to eliminate from the measured values the extortions related to the combustion process of the fuel-air mixture. In the case of measuring vibration parameters during the engine rundown, an important factor enabling the correct and precise diagnosis is the synchronization of vibration signals from the rotors with their rotational speed. This task is carried out using optical probes or magnetic sensors.

A common practice used by turbine engine manufacturers is to assign the HMS vibration module to steady loads, where the value of the acceptable vibration parameters is assigned to a specific rotor speed [1]. This highly effective approach alerts to critical problems such as unbalance and damage to bearing systems. However, the disadvantage of this solution is low sensitivity, where the alarm is triggered in an already advanced state of damage [2]. This is often due to the high engine load dynamics and the impossibility of reproducing an identical engine load in its natural operation. This results from variable resistance characteristics of the hull resulting from the variability of weather conditions, fouling of the hull, and various loading conditions of the warship. An alternative is to study the vibration parameters in the engine coasting process, i.e. in a nonstationary state. During stops, the fuel supply to the injectors is cut off, thanks to which the stopping process is not disturbed by vibrations caused by fuel combustion in the combustion chambers. Such a test allows for revealing slight changes in vibration symptoms, which may initially be assumed as vibration patterns. Importantly, the process of stopping the engine always takes place in very similar conditions and is forced by the stopping procedures required by the engine manufacturers. In addition, the rotor system of a marine turbine engine is a supercritical system, which means that during the rundown process passes through at least one resonance frequency. Measurements of vibration parameters in the resonance area can be useful in determining the technical condition of the engine. The article proposes a technique for diagnosing the turbine rotor system using the nonstationary state during the rundown of the rotor system with the use of signal synchronization with the shaft rotational speed.

Tests of bearing systems of rotating machines use typical vibration analysis resulting from the knowledge of the geometry and kinematics of rolling bearings and highly advanced signal processing tools. Cui et al., proposed for nonstationary signals a method called multivariate complex variational mode decomposition (MCVMD) [2]. For the visualization goal of HMS, they used the three-dimensional instantaneous orbit map (3D-IOM), which enables more features of shaft vibration of a rotor system to be displayed. In addition, the proposed method provides a high-resolution Time-Full Spectrum (Time-FS) to display the forward and backward frequency components of multiple bearing sections.

The literature gives some possible techniques that may be adopted in the diagnostic process of marine turbine engines. He et al. conducted diagnostic tests of wind turbine rotor systems [3]. The research has achieved a positive result after the oscillation-based decomposition of vibration signals. The low oscillatory component of a vibration signal has been separated from the original signal with the help of a Tunable Q factor Wavelet

Transform (TQWT) based on its ability to perform continuously under varying speed conditions. Then, health monitoring of the rolling element bearing was conducted by quantifying the extracted low oscillatory vibration signal component with four prominent sparsity indices: kurtosis, Gini index, negative entropy, and reciprocal smoothness index. Similar research was provided Zimroz et al. [4]. They proposed to use novel features for the long-term monitoring of wind turbines. It was shown that regression analysis parameters could be used as unvarying, fault-sensitive features for decision-making.

Zhen et al. realized the accurate fault detection of rolling element bearings based on nonstationary vibration signal analysis using Weighted Average Ensemble Empirical Mode Decomposition (WAEEMD) and Modulation Signal Bispectrum (MSB) [5]. In addition, MSB based on the modulation characteristics of the signals was developed for demodulation and noise reduction. Ensemble Empirical Mode Decomposition (EEMD) is an advanced non-linear and nonstationary signal processing approach that can decompose the signal into a list of stationary intrinsic mode functions (IMFs). The proposed method took advantage of WAEEMD and MSB for bearing fault diagnosis based on vibration signature analysis. Firstly, this method decomposes the vibration signal into IMFs with a different frequency band using EEMD. Then, the IMF's are reconstructed into a new signal by the weighted average method, called WAEEMD, based on Teager Energy Kurtosis (TEK). Finally, MSB is applied to decompose the modulated components in the reconstructed signal and extract the characteristic fault frequencies for fault detection. The experimental results verified that the WAEEMD-MSB has superior performance over conventional MSB and EEMD-MSB in removing fault features for rolling element-bearing fault detection.

Wu et al. analyzed the vibration signals extracted from the variable speed that contain more fault information than those operating in steady-state conditions [6]. To solve problems of missing harmonics, a proposed method was developed based on improved Variational Mode Decomposition (VMD) by Generalized Demodulation Technology (GDT) and a Zero-Phase Shift Filter (ZPSF) and adopted for the startup process of the rotor machine. The authors noticed that the proposed method could effectively compromise the speed fluctuation and thus extract all the helpful vibration components from the rotor system during startup, and it is an effective tool for fault diagnosis.

Regarding the gas turbine diagnostics test, Krzyworzeka et al. advanced method - Procedure of Linear Decimation (PLD), was applied for signal analysis of vibrations of GTD-350 turbine engine in unstable operating conditions [7]. The rotational speed of the driving shaft was adopted as a reference signal. The results were covered for turbine running up and coasting state. Many authors note the benefits of analyzing the vibration parameters of the rotor systems recorded during operation in the starting and stopping conditions. Czyż [8] studied loads acting on an unmanned helicopter's drive unit during acceleration and braking maneuvers, objects of the investigation were PZL W3-Falcon helicopter engines. Also, Kilikevičius et al. [9] focused on flying objects analyzing the influence of parameters of aircraft propeller modal parameters on small airplane performance. Zachwieja in his work on rotor systems, slightly different rotor systems than considered in this article, because they work in industrial pumps [10], show the possibility of using resonance parameters for effective diagnostics of their technical condition.

All indicated studies focused on applying procedures in online type technical diagnostics [11]. In operational practice, there is a significant population of older turbine engines for which online systems use is not economical. In the case of such engines, it is often impossible to use modern stationary diagnostic systems due to the need to install many sensors in an invasive way. Users often do not agree to interference in the structure of the engine. The exploitation of such machines is most often carried out in the time-schedule procedure. It means the risk of not noticing early failure symptoms or performing repairs on technically efficient engines. Damage to turbo engine components often propagates relatively slowly and it is possible to detect them during periodic measurements. In this work, the authors proposed an offline vibration diagnostics method to clearly indicate changes in the technical condition of bearings in the rotor systems of a turbine engine. The research aimed to prepare a diagnostic procedure for the rundown of the rotor system and determine the value of vibration symptoms of the pattern type and the most sensitive measurement sites. The conducted research aimed to apply simple and cheap vibration recorders and to prepare a diagnostic path for technical personnel with basic diagnostic knowledge.

2 Material and Methods

The tests were carried out on four LM2500 - see Fig. 1. The Reflex system by Bruel & Kjaer was used for measurements and analyses. Accelerometers were mounted on steel struts bolted above the middle bearing of compressor rotors. The triggering signals were obtained from the optical probe permanently installed in the gas turbine LM2500. Tests were performed for the idle load rotor shutdown process after 3 min of gas turbine cooling. Before and after the diagnostic tests were performed, modal hammer tests were carried out to identify local resonances. The primary criterion for the stopping process was compliance of the planned gas generator (GG) rotational speeds, the temperature of exhaust gases and the air temperature in the inlet duct.

Fig. 1. Accelerometer mounting location over the middle bearing of compressor of the LM2500 Gas Turbine.

The criterion identifying changes in the technical state of the bearing system and the mass structure of the rotor system of the exhaust gas generator of the LM 25000 engines are changes in the values of the resonance frequencies. That might be detected using the Order Tracking procedure in the rotor rundown process. The research identified the characteristic features of the vibration velocity amplitude in the 1st order range (synchronous harmonic) in the frequency domain (converted to the rotor speed in the deceleration process in Hz). This analysis allows us to identify in detail the engine rundown process:

- from nGG = 5000 rpm to transient speeds nGG = 3000 rpm,
- from nGG = 5000 rpm to transient speeds nGG = 2000 rpm,
- from nGG = 5000 rpm to transient speeds nGG = 1000 rpm.

The diagnostic tests concerned the following vibration symptoms identifying changes in the technical condition of the rolling bearings of the gas turbine rotor system:

- changes in the rundown time between idle speeds and control speeds,
- analysis of changes in the first order value for vibration velocity during deceleration,
- detection of the appearance of new resonance components during the Order Tracking analysis [12, 13].

The rundown time for the tested turbines had its reference values determined during the first tests after all engines overhaul. The differences in the coast-down time for the successively tested engines did not exceed the value of 4 s for the analyzes from $n_{GG} =$ 5000 rpm to speed $n_{GG} = 1000$ rpm. The tests were conducted on average every 150 h of engine operation, depending on the operational possibilities of the ships.

The analysis of changes in the values of the first orders required the execution of model patterns for individual engines and the identification of 1st order resonances as a function of rotational speed and time. In addition, changes in the speed/frequency domain identified changes in rotor stiffness.

In the rotor deceleration process, new resonances may appear, not necessarily related to the rotor system. Their identification is facilitated by Order Tracking analysis, where the resonance can be identified in the time and frequency domains. The FEM model of the rotor and numerical simulations allows parts of turbine resonance identification using inverse models.

3 Analysis of Vibration Signals in Nonstationary States

The analysis of vibration parameters in transient states is a diagnostic tool commonly used not only in the operation of turbine engines but also in mining, wind farms, and the automotive and chemical industries [14]. Variable load states and a wide spectrum of operating rotational speeds or their non-stationarity require the analysis of vibration signals using synchronizing signals [15]. These analyses are carried out in many ways; however, all of them are aimed at the pseudo-stationarity of the analysed signal in the domain of time or rotational speed [16]. The most common methods of analysing vibration parameters measured in transient states are:

- synchronism of the signal in the field of rotational speed by triggering one or more synchronizing signals per one revolution of the machine shaft,

- oversampling of the signal [7],
- order tracking [17],
- order autotracking [17].

The most common way to acquire a synchronizing signal is by the following methods:

- optical sensor,
- laser sensor,
- tachometric generator,
- inductive sensor.

The paper presents the results of tests carried out using signal synchronism and order tracking when acquiring the synchronizing signal from turbine engine rotors with the optical method (LM 2500). However quite often it is not possible to measure velocities of rotors in turbine engine. In such situations velocity curve might be obtained from vibration signal with the use of auto tracking procedure. The frequency of the 1st order harmonic amplification in the deceleration process is characteristic of an efficient rotor-pattern symptom (see Figs. 2 and 3).

In the analysed FFT spectrum, it is possible to detect new components in the deceleration process that have not been seen so far (Fig. 2). The graph shown in Fig. 3 allows us to determine the deceleration time between idle load $n_{GG} = 5000$ rpm with an accuracy of 0.1 s and 5 rpm to the subsequent control GG rotor speeds.

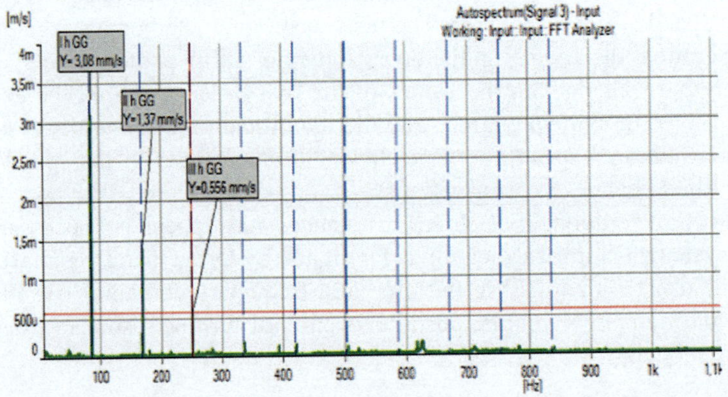

Fig. 2. Examples of parameters during idle load of the LM 2500 type GT - vibration velocity spectrum for nGG = 5000 rpm.

Direct measurement of the rotational speed of individual elements of the marine turbine rotor system is sometimes difficult or even impossible. In such cases, autotracking procedure may be helpful. The use of this method enables the extraction of the rotational speed signal from the vibration signal. It is necessary to know the approximate value of the first harmonic frequency for any steady state (e.g. start of the engine stop procedure - idling). The signal processing literature describes the problem of detecting periodic signals as calculating the amplitude and phase of several harmonic components [17]:

$$d_m(t) = a_{m,0} + [\sum_{k=1}^{K} a_{m,k}\cos(\omega_k t) + b_{m,k}\sin(\omega_k t)] + e_m(t) \quad (1)$$

where:

$e_m(t)$ – noise term with a mean value equal to zero,

a_k – next orders of the fundamental frequency.

When considering a series of samples (Eq. 1), it can be formulated as a linear problem [17]:

$$\boldsymbol{d} = \boldsymbol{G}\boldsymbol{b} + \boldsymbol{e} \tag{2}$$

where:

$\boldsymbol{d}$ – column vector of measured data,
$\boldsymbol{G}$ – matrix represents cosines and sines,
$\boldsymbol{b}$ – column vector of amplitudes,
$\boldsymbol{e}$ – noise contribution.

Based on Bayes's thesis, the joint posterior probability density function (PDF) of the parameters in a model is equal to the possibility of the data multiplied by the earlier knowledge of the parameters divided by the evidence of the data. It can be written as:

$$P(A/\boldsymbol{d}) = \frac{P(\boldsymbol{d}/A)P(A)}{P(\boldsymbol{d})} \tag{3}$$

where:
A – model parameters;
$\boldsymbol{d}$ – data parameters.

So, it is possible to calculate the posterior distribution of ω_0 conditioned on the measurement data. The use of an autotracker allows the calculation of a series of probability spectra, giving the probability that a given frequency illustrates the RPM. Based on other statistics, the final rotational speed profile is estimated. Using the procedure must provide some input to the algorithm that determines the instantaneous rotational speed; the main ones are the actual RPM range, the maximum acceleration [rpm] and the dominant RPM orders/harmonics in the example of a vibration signal. A necessary condition is usually the determination by the user of the expected frequencies dominating the spectrum and linking them with the kinematics of the tested device. In addition, the analysis of the rundown (non-linear) processes must provide information about the maximum rotational speed expected in the signal. Changes in the rundown time and the transition time between rotational speeds specific to a given rotor system usually provide us with information about the condition of the bearing system.

Time gains are due to rolling bearing wear or leaks between the compressor blade disc, seal and spacer ring. In the case of the engines being the subject of the tests, this symptom is most visible in the range from $n_{GG} = 5000$ rpm to the speed of $n_{GG} = 1000$ rpm, which results from carbon deposits on the rotor shaft and increased rotational resistance. Furthermore, identification of changes in the stiffness of the rotor system is possible through the analysis of orders and, above all, the title of skew components in the deceleration process – see Fig. 4. The tests were carried out after performing a routine inspection, one of the elements of which is washing the engine compressor. This procedure allowed the rotor system to be cleaned of most carbon and salt impurities. These activities ensured the restoration of the correct weight distribution of the rotors relative to their axis of rotation. Removal of deposits meant that they do not affect the braking time of the rotors. Failure or damage to other engine components or ancillary equipment in the housing module may occur if gas turbines are under normal operation. Typical diagnostic systems are programmed for specific symptoms of the technical condition of the rotor system; therefore, these cases can be omitted under operating conditions. For example, the Order Tracking procedure identifies changes in dynamic vibration parameters during the rotor deceleration process. The example of Order Tracking shown in Fig. 4 shows the lack of unusual local resonances both in the rotor part and in the frequency range up to the 10th order of the GG rotor.

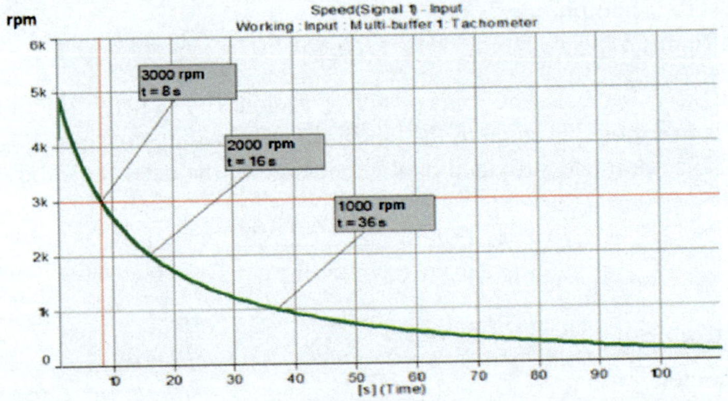

Fig. 3. Examples of parameters during nonstationary states tests for LM 2500 – a graph of rotational speed changes of the GG rotor from the idle load in the rundown process.

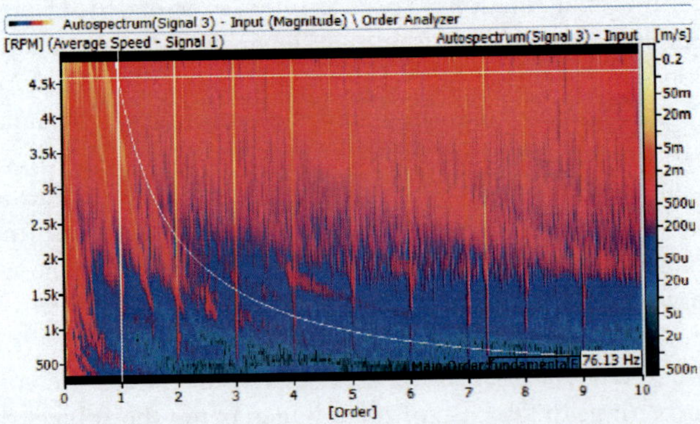

Fig. 4. Analysis of the orders (vibration velocity) in the deceleration process of the GG rotor.

The analysis of the first three orders in the function of time and rotational speed of the GG rotor is an effective diagnostic tool - see Figs. 5 and 6. Both graphs represent the individual vibration fingerprint of each rotor. Observation of changes in local resonances regarding their value and migration in the rpm domain (frequency) provides valuable information on the rotor-bearing system. Local resonances are described by the time of occurrence, the rotational speed of the rotor and the amplitude of the vibration velocity I-st order of the GG rotor. The temperature of the lubricating oil is an essential factor influencing the quality of diagnostic test results. It was noted in the research that this factor could significantly affect the deceleration time, accelerate the time of occurrence of the phenomena and increase the amplitudes of local resonances. Hence, all tests were carried out with a tolerance of $\pm 5^0$C of oil temperature at the inlet to the engine.

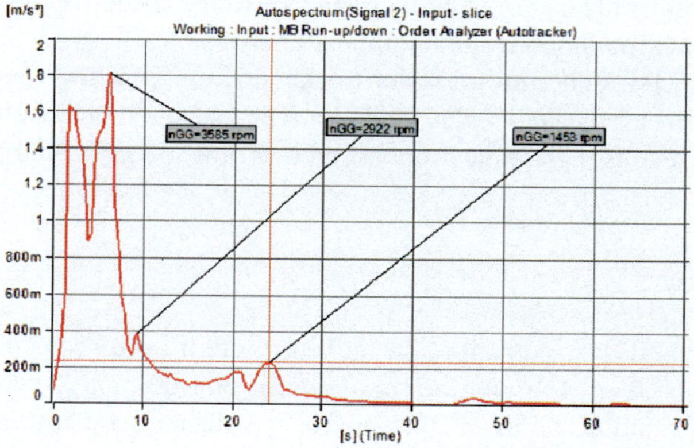

Fig. 5. An exemplary analysis of the 1st order tracking of the GG rotor of the LM 2500.

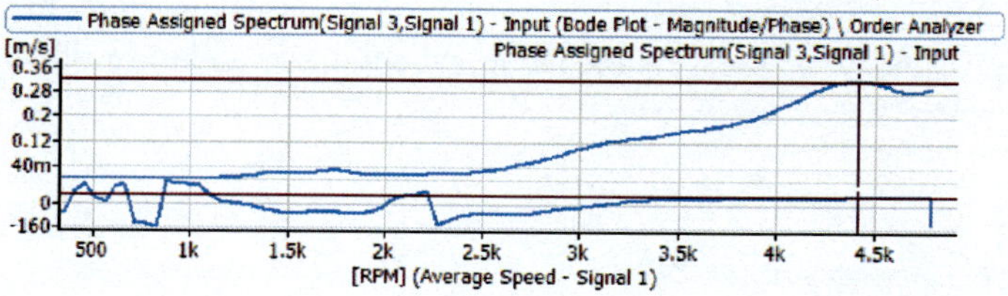

Fig. 6. An exemplary analysis of the 1st order tracking of the GG rotor of the LM 2500 as a function of the coast-down rotation of the GG rotor.

4 Discussion

The presented methodology and research results are used to assess the technical condition and forecast changes in the bearing arrangements of GG rotors of the LM2500 type gas turbines. The technical condition assessment method consists of three steps, i.e.:

1. Determining the GG rotor rundown time for three rotational speed ranges and comparing them with the reference values for reference measurements on each rotor. The time tolerances for the intervals are as follows:

   - from $n_{GG} = 5000$ rpm to speeds $n_{GG} = 3000$ rpm - 3 s,
   - from $n_{GG} = 5000$ rpm to speeds $n_{GG} = 2000$ rpm - 4 s,

- from $n_{GG} = 5000$ rpm to speeds $n_{GG} = 1000$ rpm - 6 s.

Observation of the shortening of the coasting time means wear of the bearing or the formation of carbon deposits, increasing frictional resistance.

2. Analysis of the Orders (vibration velocity) in the deceleration process of the GG rotor to identify resonances. In the studied spectrum, deviations from the recorded results from the standard rundown measurement with this method are sought. Identifying potential failures or resonances involves analysis in the speed/frequency domain and comparing FEM simulation results in modal analysis.
3. Analysis of the 1st, 2nd and 3rd order trackers of the GG rotor and evaluation of changes in the occurrence of local resonances. The measurement results are compared with reference results in the time domain and the rotor speed GG during the rundown process.

5 Conclusions

The presented method is an element of the multi-symptom gas turbine diagnostic system used by the Polish Navy. The vibration module for nonstationary states is offline and is used ad hoc, depending on the availability of gas turbines for testing. A significant advantage of the proposed method is obtaining information on the condition of the bearing system, which is not the subject of online research. Online monitoring systems are dedicated to symptoms of unbalancing and misalignment problems. Thus, the presented diagnostic module complements the highly effective diagnostic system. Furthermore, it increases the possibility of assessing the technical condition and wear of bearings in the rotor system.

References

1. Matuszczak, M., Żbikowski, M., Teodorczyk, A.: Predictive modelling of turbofan engine components condition using machine and deep learning methods. Eksploatacja i Niezawodność – Maint. Reliabil. **23**(2), 359–370 (2021). https://doi.org/10.17531/ein.2021.2.16
2. Cui, X., Huang, J., Li, C., Zhao, Y.: Three-dimensional instantaneous orbit map for rotor-bearing system based on a novel multivariate complex variational mode decomposition algorithm. Mech. Syst. Signal Process. **178**, 109210 (2022). https://doi.org/10.1016/j.ymssp.2022.109211
3. He, W.P., Zi, Y.Y., Chen, B.Q., Wang, S., He, ZhengJia: Tunable Q-factor wavelet transform denoising with neighboring coefficients and its application to rotating machinery fault diagnosis. Sci. China Technol. Sci. **56**(8), 1956–1965 (2013). https://doi.org/10.1007/s11431-013-5271-9
4. Zimroz, R., Bartelmus, W., Barszcz, T., Urbanek, J.: Diagnostics of bearings in presence of strong operating conditions non-stationarity—a procedure of load-dependent features processing with application to wind turbine bearings. Mech. Syst. Signal Process. **46**(1), 16–27 (2014). https://doi.org/10.1016/j.ymssp.2013.09.010
5. Zhen, D., Guo, J., Xu, Y., Zhang, H., Gu, F.: A novel fault detection method for rolling bearings based on non-stationary vibration signature analysis. Sensors **19**(18), 3994 (2019). https://doi.org/10.3390/s19183994

6. Jili, W., Zhang, X., Li, B.: A study on vibration component separation of a rotor system during startup and its application in fault diagnosis. Meas. Sci. Technol. **30**(9), 095104 (2019). https://doi.org/10.1088/1361-6501/ab199c
7. Krzyworzeka, P., Cioch, W.: Demodulation of nonstationary machine vibration using cycle-time scale. Arch. Acoust. **31**(2), 167–177 (2006)
8. Czyż, Z., Magryta, P.: Analysis of the operating load of foil-air bearings in the gas generator of the turbine engine during the acceleration and deceleration maneuver. Eksploatacja i Niezawodnosc – Maint. Reliabil. **18**(4), 507–513 (2016). https://doi.org/10.17531/ein.2016.4.5
9. Kilikevičius, A., Rimša, V., Rucki, M.: Investigation of influence of aircraft propeller modal parameters on small airplane performance. Eksploatacja i Niezawodność – Maint. Reliabil. **22**(1), 1–5 (2020). https://doi.org/10.17531/ein.2020.1.1
10. Zachwieja, J.: Effectiveness of diagnosing damage to an industrial pump rotor by analysing its vibrations. Diagnostyka **20**(1), 33–39 (2018). https://doi.org/10.29354/diag/99604
11. Kan, M.S., Tan, A.C.C., Mathew, J.: A review on prognostic techniques for non-stationary and non-linear rotating systems. Mech. Syst. Signal Process. **62–63**, 1–20 (2015). https://doi.org/10.1016/j.ymssp.2015.02.016
12. Grządziela, A., Kluczyk, M.: An application of order tracking procedure for diagnosis technical state of rotor system in shut-down process. J. KONES. Powertrain Transp. **20**(1), 111–118 (2013). https://doi.org/10.5604/12314005.1135323
13. Grządziela, A., Kluczyk, M.: The impact estimation of damping foundations in dynamics of the rotor system. Diagnostyka **18**(3), 7 (2017)
14. Dziurdź, J.: Transformation of nonstationary signals into pseudostationary signals for the needs of vehicle diagnostics. Acta Physica Polonica A **118**(1), 49–53 (2010). https://doi.org/10.12693/APhysPolA.118.49
15. Bendat, J.S., Piersol, A.G.: Random Data: Analysis and Measurement Procedures: Fourth Edition. Random Data: Analysis and Measurement Procedures, 4th edn. (2012). ISBN: 978-1-118-21082-6
16. Noman, K., Li, Y., Peng, Z., Wang, S.: Continuous health monitoring of bearing by oscillatory sparsity indices under non stationary time varying speed condition. IEEE Sensors J. **22**(5), 4452–4462 (2022). https://doi.org/10.1109/JSEN.2022.3146264
17. Pedersen, T., Herlufsen, H., Hansen, H.: Order tracking in vibro-acoustic measurements:a novel approach eliminating the tacho probe. SAE Technical Paper 2005–01–2266 (2005). https://doi.org/10.4271/2005-01-2266

Theoretical-Experimental Determination of CVT Power Losses Due to Rubber V-belt Vibrations

Waldemar Łatas[1] and Adam Kot[2](✉)

[1] Department of Applied Mechanics and Biomechanics, Faculty of Mechanical Engineering, Cracow University of Technology, Kraków, Poland
waldemar.latas@pk.edu.pl

[2] Department of Automotive Vehicles, Faculty of Mechanical Engineering, Cracow University of Technology, Kraków, Poland
adam.kot@pk.edu.pl

Abstract. The article proposes a solution of determining power losses caused only by rubber V-belt vibrations in continuously variable transmission (CVT). The analysis was performed on the basis of a mathematical model and experimental data. The article includes the results of experimental tests of vibrations of a transmission belt driven by a two-stroke engine. A contactless measurement was performed by a high-speed camera and a time-lapse image analysis was performed in dedicated software. This new approach in the discussed research area allows obtaining results impossible to receive with other measurement methods. The mathematical model applied is an axially moving, with constant velocity, Euler-Bernoulli beam, the nonlinear effects and longitudinal vibrations are ignored. It was assumed that the material of the belt obeyed the Voigt-Kelvin rheological model. The method of determining the coefficients describing the damping properties of the belt is also discussed. The equivalent bending damping coefficient was determined based on the analysis of the damped free vibrations of the cantilever beam, made of the belt segment. The CVT power losses for fixed working conditions were obtained after numerical calculations with the use of procedures of the IMSL® C package.

Keywords: Continuously Variable Transmission · Moving Beam Vibrations · High-speed Camera · Power Losses

1 Introduction

For many years the belt drive has been used widely in vehicle drive systems. In modern cars stepless gearboxes with steel or hybrid (rubberized steel) push belts have most common use. In small vehicles (e.g. scooters, snowmobiles) the powertrain usually uses rubber V-belt. A stepless gearbox, called continuously variable transmission (CVT), allows a much better use of the power characteristics of the internal combustion engine, where for every power level there is one-speed-torque combination which achieves

A. Puchalski et al. (Eds.): ICTD 2022, ACM 21, pp. 32–41, 2023.
https://doi.org/10.1007/978-3-031-31719-4_4

optimal fuel efficiency. Using a continuously variable range of transmission ratio, a line connecting the operating points can be followed by high drive unit efficiency [1, 2]. In this approach, the advanced transmission ratio controller (electromechanical or electrohydraulic) is needed. On the other hand, in two-wheelers with rubber V-belt CVT only simple mechanical regulator is used. The main advantages of CVT application in these kinds of vehicles are powertrain simplicity, reliability and low price. The mentioned reasons the vehicles like scooters are especially commonly used in south-east Asia. For example, this sector determines more than 80% of the market for vehicles in India [3]. The main disadvantage of the transmissions with rubber belt results from the lower efficiency. The power losses usually reach from 20% to 50% [4]. It is a very high value in comparison with the classic gear transmission based on toothed gears where efficiency is higher than 95%.

There are two major methods of improving the efficiency of rubber belt CVT. The first, as mentioned before, is combined with advanced transmission ratio regulation methods. Such a solution entails some difficulties such as cost and huge powertrain complexity. The second way of efficiency improvement refers to the friction pairs' (belt and pulleys) properties and the belt behavior. The complexity of this issue results from, e.g. the V-shape cross section of the belt, the changes of forces acting on the belt in one driving cycle, the belt transverse and longitudinal flexibility, the rubber hysteresis [5–7].

The vibration issue in belt transmission has also been the subject of research by many authors [8–13] however, these were mainly the theoretical investigations. There are two basic approaches in the description of belt dynamics in the literature. If flexural rigidity of the axially moving system can be neglected, the string model is used (e.g. narrow and long belts) [14–16]. Otherwise, one of the beam theories (Euler-Bernoulli, Timoshenko or Rayleigh) is applied [17, 18]. The works [19, 20] include the comprehensive reviews of different mathematical models, both linear and nonlinear.

It should be strongly emphasized that the problem of the influence of belt vibrations on power losses is an unexplored area both in theoretical and experimental terms. Although there are extensive works on CVT power losses in the literature [21, 22], belt vibrations are not taken into account.

This work presents the research approach and measurement results of the belt transverse vibrations obtained with the use of a high-speed camera. The showed test results reflect very well the actual working conditions of the considered CVT because the considered transmission is powered by a two-stroke single-cylinder internal combustion engine, which is a standard source of drive for scooters. The paper proposes a theoretical approach to power losses determination on the basis of presented experimental data. The method of determining the coefficients describing the damping properties of the belt is also discussed.

2 Test Stand and Research Software

The registration of the CVT belt transverse vibrations were conducted on a test stand equipped with complete transmission of the TGB 101S scooter, driven by a two-stroke engine with a cubic capacity 49 cm^3. The drive unit develops 3.5 kW of power. The diagram of the test stand is presented in Fig. 1. Such a configuration of the test stand allows measuring in real working conditions.

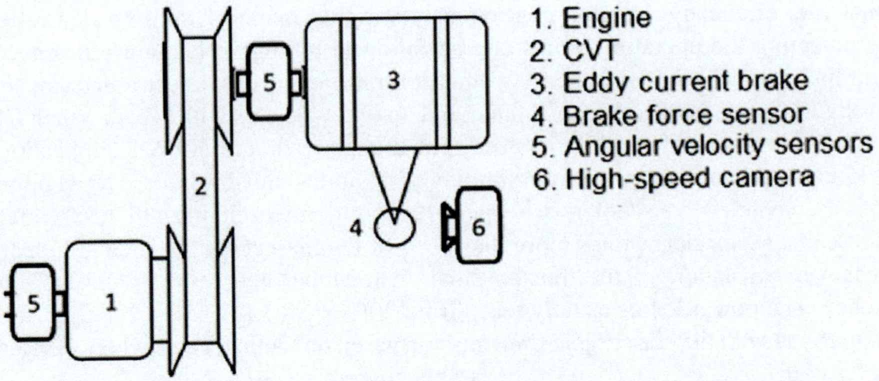

Fig. 1. Test stand scheme.

Fig. 2. View of test stand with high-speed camera.

The application of a high-speed camera in the analysis of belt movement is a new approach proposed e.g. in the work [23]. The belt transverse vibrations were registered by a high-speed camera Phantom VEO 710S (Fig. 2). The footage was recorded at the 1280 × 720 resolution with the frequency of 8300 fps. At the rotational speed of 6000 rev/min, such a frequency allows registering an image every 4.3° of the rotation angle of the crankshaft. Time-lapse image analysis was performed by TEMA Motion 4.2 software. The program allows defining own coordinate system and saving the coordinates of selected points. The belt tracks, for both tight and slack sides were registered by selecting the points equally spaced from each other (14 points on each belt side). The time-lapse analyses were conducted for single crankshaft revolution in fixed working conditions defined by throttle position (100%), engine revs (three values considered: 3150, 4560 and 6000 rev/min) and transmission ratio (0.9).

3 Physical Structure of Transmission Belt

Due to the non-homogeneity of a given cross-section (Fig. 3) and the variability of the cross-sectional area along the length of the belt, theoretical determination of values of the physical parameters describing the behavior of the moving vibrating transmission belt would be very complex.

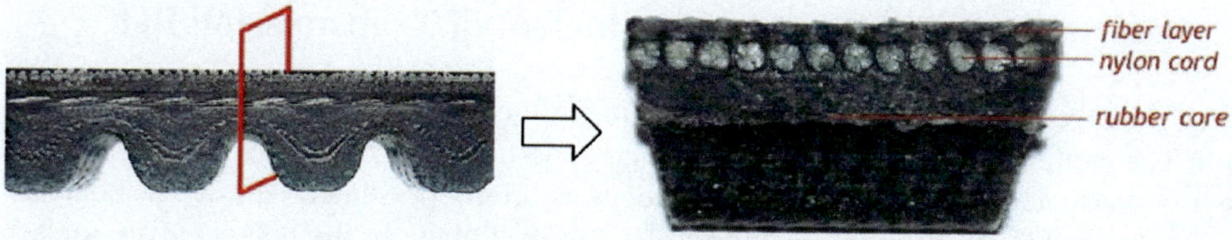

Fig. 3. Cross section of CVT rubber V-belt [DAYCO 17.5 × 765].

Numerical determination of these values, for example with the use of FEM, would also be time consuming. Both methods would require knowledge of the material constants of each of the belt components, i.e. nylon cord, rubber core and fiber layer. Additionally, it would be necessary to know the way these elements interact with each other, which would in turn demand expertise of the production process technology. The observations showed that in spite of the heterogeneous belt structure, small vibrations of the cantilever beam (made of the transmission belt section) disappear with a very good approximation in accordance with the damping described by the linear Voigt-Kelvin rheological model. Despite various phenomena occurring in the axially moving and vibrating transmission belt which cause energy dissipation, their quantitative description using one equivalent factor is therefore fully justified.

4 Mathematical Model of Transmission Belt Vibrations

The mathematical model used in the test is an axially moving, with constant velocity, Euler-Bernoulli beam, whereby nonlinear effects and longitudinal vibrations are neglected. It was assumed that the material of the belt conformed with the Voigt-Kelvin rheological model, with the constitutive stress-strain relation given by:

$$\sigma = E\varepsilon + \alpha(\varepsilon_t + V\varepsilon_x), \tag{1}$$

where E and α are Young's modulus and the viscous damping coefficient, V is the transport velocity of the axially moving belt. An expression for the bending moment can be obtained from formula (1):

$$M = \overline{EI}w_{xx} + \overline{\alpha I}(w_{xxt} + Vw_{xxx}). \tag{2}$$

Forms of expressions (1) and (2) stemmed from the material time derivative applied. Due to the belt being non-homogeneous, the equivalent coefficients (averaged along the belt length) are introduced: $\overline{\rho A}$, $\overline{EI}$, $\overline{\alpha I}$, where ρ represents density, A and I are cross-section area and area moment of inertia, respectively. For the mentioned assumptions the equation of the transverse motion of the moving belt is given by:

$$\overline{\rho A}\left(w_{tt}+2Vw_{xt}+V^2w_{xx}\right)-T_0w_{xx}+\overline{EI}w_{xxxx}+\overline{\alpha I}(w_{xxxxt}+Vw_{xxxxx})=0, \quad (3)$$

where $w(x,t)$ indicates the transverse displacement, x and t represent the axial and time coordinates, T_0 is the static belt tension.

5 Determination of Damping Coefficient of Transmission Belt

Equation (3) contains several parameters having influence on the vibrations of the moving belt. The method of determining the equivalent coefficient $\overline{\alpha I}$, on which the value of the energy dissipated due to bending is dependent, is briefly presented below. The analysis of the damped free vibrations of the cantilever beam, made of the belt segment, makes it possible to determine the equivalent bending damping coefficient $\overline{\alpha I}$. The vibration measuring stand is shown in Fig. 4 – in order to avoid the effect of gravity on the deflection line, the vibrations of the belt occur in the horizontal plane. The measurement was performed by means of two (for verification) optical distance sensors Baumer FADK 14U4470/S14/IO (measurement range: 50 ÷ 400 mm, accuracy: 0.1 mm).

The determination of the viscous damping coefficient $2h$ of the equivalent vibrating discrete system with one degree of freedom provides a basis for the calculation of the coefficient $\overline{\alpha I}$, which is described by the equation of motion:

$$m\ddot{y}+2h\dot{y}+ky=0, \quad (4)$$

where y denotes the displacement of the beam end point. The mode in which the beam vibrates was assumed as the line of static deflection of a cantilever beam under a concentrated force applied at its end. Compared kinetic energies of the beam and the equivalent 1.d.o.f. system give:

$$m=\frac{33}{140}m_b, \quad (5)$$

where m_b is the mass of vibrating section of the belt with a length of $l=0.150\,\mathrm{m}$, equal to: $m_b=0.020\,\mathrm{kg}$.

In the plots in Fig. 5 examples of displacement waveforms of the beam end point are demonstrated.

Based on the presented waveforms, the logarithmic decrement of damping was determined, subsequent to simple calculations the following value of the coefficient h was received:

$$h=0.022\frac{\mathrm{kg}}{\mathrm{s}}. \quad (6)$$

Comparing energy dissipations in the beam and in the damper of the 1.d.o.f equivalent system (during one vibration period):

$$\overline{\alpha I}\int_0^l\int_0^T(w_{xxt})^2dxdt=2h\int_0^T\dot{y}^2dt \quad (7)$$

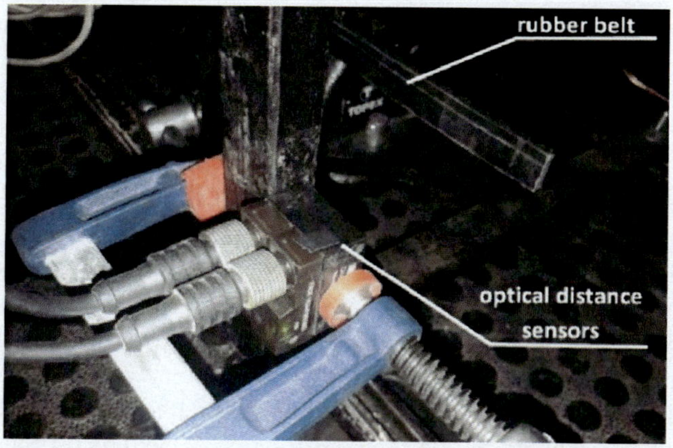

Fig. 4. Transmission belt segment vibration measurement stand.

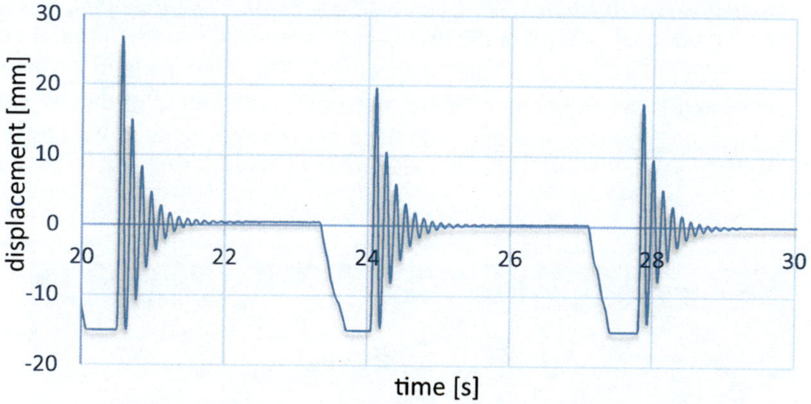

Fig. 5. Transverse displacement waveforms measured at end point of belt segment.

the following formula is obtained:

$$\overline{\alpha I} = \left(\frac{35}{214}\right) l^3 h = 1.214 \cdot 10^{-5} \frac{\text{kgm}^3}{\text{s}}. \quad (8)$$

6 Determination of Power Dissipated During Transmission Operation

The average power D of the energy dissipated due to bending in the moving transmission belt on the section $[l_1, l_2]$ in the time interval $[t_1, t_2]$ is provided by the formula:

$$D = \frac{1}{t_2 - t_1} \int_{l_1}^{l_2} \int_{t_1}^{t_2} \overline{\alpha I} (w_{xxt} + V w_{xxx})^2 dx dt. \quad (9)$$

Figure 6 shows the shape of the belt while the transmission is in operation at a given moment of time, with marked points, the positions of which (being distances from the horizontal axis) are defined by means of the high-speed camera. The intervals between the projections of these points on the horizontal axis are equal to 9.32 mm. It should be noted with an emphasis that in the following moments of time, as a result of the axial movement, the points indicate physically different locations of the belt. Based on the measurements, the matrix of positions $\left[\tilde{w}_{ij}\right]$ is defined, where $\tilde{w}_{ij} = \tilde{w}(\tilde{x}_i, t_j)$ indicates the position of the point with the coordinate $\tilde{x}_i$ at the time t_j. The positions $\tilde{w}_{ij}$ and the $\tilde{x}_i$ coordinates are transformed (by rotation) into a system with its abscissa axis parallel to the tangent to the transmission pulleys. In such coordinate system, interpolation of the w_{xxt} and w_{xxx} derivatives is performed for the new positions w_{ij} and the x_i coordinates, making it possible for the dissipated power to be calculated according to the formula (9). In the numerical calculations, for the interpolation as well as for the calculation of the double integral, the numerical procedures of the IMSL® C package were employed.

The average values of power dissipated in the transmission belt for rotational speed of the drive pulley are outlined in Table 1 and equal to: 3150, 4560 and 6000 rev/min, for both the tight and the slack sides. For the listed rotational speed values, the axial velocity V values are: 13.96, 20.21 and 26.59 m/s, respectively. The time in which the dissipated energy is calculated is synonymous with the time of one revolution of the drive pulley. In all cases, the same section of the belt with the length of approximately $\Delta l = 0.121$m were covered by the calculations. The dimensions of matrix $\left[w_{ij}\right]$ are 14 × 23.

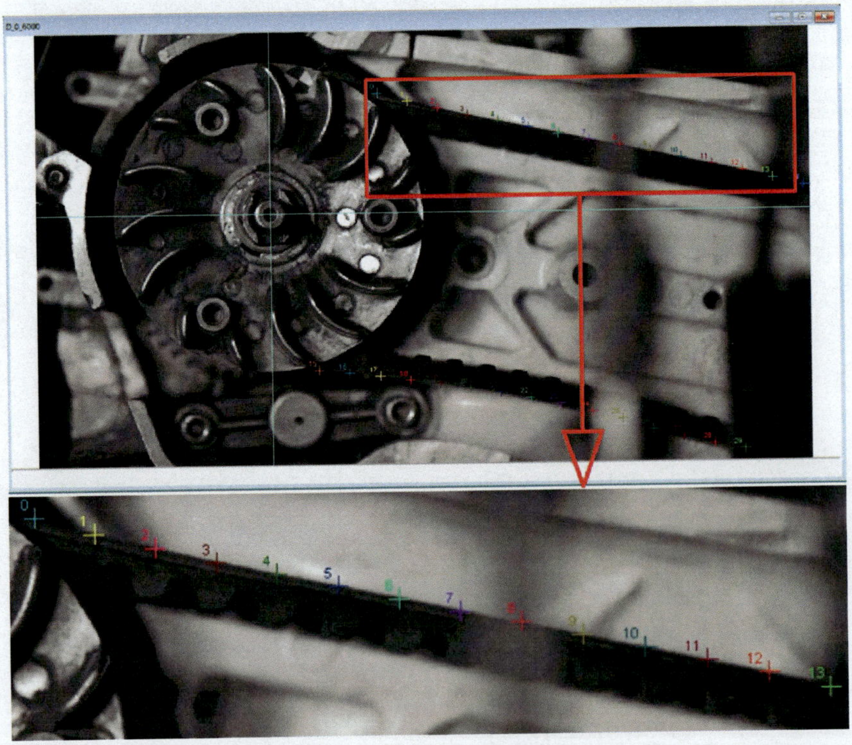

Fig. 6. Belt vibrations – measurement points. View in TEMA Motion 4.2 software.

Table 1. Average power (in Watts) dissipated in transmission belt for various rotational speeds.

	n= 3150 rev/min	n= 4560 rev/min	n= 6000 rev/min
Tight belt side (upper fragment)	4	27	10
Slack belt side (lower fragment)	39	31	113
Sum	**43**	**58**	**123**

7 Discussion

The values of power losses presented in the article relate to only the segment of the belt captured by the high-speed camera. About a quarter of the belt was left out of the frame (covered up by eddy current brake). However, it should be emphasized that the obtained values correspond with efficiency of the an identical CVT, where overall power loss (taking belt slip, hysteresis, etc. into consideration) in similar working conditions is in the range 500–1000 W [4]. Thus, approximately 10–15% of the power loss occurs as a result of belt vibrations. It is extremely difficult to verify of the received values due to the lack of similar analyzes in the subject literature. In general, a greater power loss is caused by the slack belt side. The similar values of the power dissipated on the tight and slack sides for $n = 4560$ rev/min are associated with the fact that this rotational speed expressed in hertz almost exactly coincides with the second natural frequency of the moving belt [23].

It should be noted with an emphasis that, except for the errors which stem from the application of the adopted physical model (Voigt-Kelvin model, Euler-Bernoulli beam), the obtained values of the dissipated power are also burdened with a large, difficult to estimate error which occurred due to the fact that determination of the third-order derivatives on the basis of discrete values of displacements obtained from measurements was needed. The displacements, based which on the functions present in the formula (9) are interpolated, are measured with 0.1 mm accuracy. Considering the errors that could possibly appear with each measurement of the displacement, a dense bundle of interpolated functions was created and the average value of the dissipated power was determined.

8 Conclusions

The power losses of rubber belt CVT generally fall into the categories: sliding losses between belt and pulleys; hysteresis of the belt material in both longitudinal and transverse directions; work required to engage and disengage the belt in and from the pulley [21]. The subject literature does not identify vibration phenomena as a factor having a significant influence on the efficiency of rubber V-belt CVT.

The presented test results are not contradictory with the approach to the causes of power losses. On the other hand, the received values of power dissipated in the belt are not negligible and prove that there is room for increasing the efficiency of this type of transmission. On the global scale, improving CVT efficiency even by several percent can make a measurable difference when fuel consumption and pollutant emission from exhaust gases are taken into account.

The obtained results demonstrate that the adopted new methodology is correct and capable of estimating power dissipated by the vibrating belt. More extensive research in this area is expedient. The losses for different belts (as different manufacturers' belts are available on the market) and for variant working conditions (transmission ratio, engine speeds) are particularly advised to be compared, which will make it possible to distinguish the parameters that could especially contribute to the minimization of power losses caused by belt vibrations.

References

1. Pfiffner, R., Guzzela, L., Onder, C.H.: Fuel – optimal control of CVT powertrains. Control. Eng. Pract. **11**, 329–336 (2003)
2. Grzegożek, W., Szczepka, M., Kot, A.: the analysis of applying CVT gear ratio rate control for scooter efficiency improvement. Asian J. Appl. Sci. Eng. **6**(2), 73–80 (2017)
3. Bhowmick, G., Sahoo, T., Bhat, A., Mathur, G., Gambhir, D.: Approach for CO2 reductioning India's automotive sector. SAE International 28-2388 (2019)
4. Grzegożek, W., Kot, A., Szczypiński-Sala, W.: The analysis of an influence of rubber V-belt physical properties on CVT efficiency. IOP Conf. Ser. Mater. Sci. Eng. **421**(2), 1–10 (2018)
5. Cammalleri, M.: A new approach to the design of speed-torque-controlled rubber V-belt Variator. Proc ImechE. **219**, 1413–1427 (2005)
6. Julió, G., Plante, J.S.: An experimentally – validated model of rubber-belt CVT mechanic. Mech. Mach. Theory. **46**, 1037–1053 (2011)
7. Gerbert, G.B.: Belt slip-a unified approach. J Mech Des. **118**, 432–438 (1996)
8. Andrianov, I.V., Horrsen, W.T.: On the transversal vibrations of a conveyor belt: applicability of simplified models. J. Sound Vib. **31**, 822–829 (2008)
9. Ferguson, N., Lin, H., Chen, E.: Experimental investigation of the transverse nonlinear vibration of an axially travelling belt. J. Vibroeng. **18**(8), 4885–4900 (2016)
10. Zhang, R., Si, X., Yang, W., Wang, N.: Analysis of resonance reliability for synchronous belt transmission with transverse vibration. J. Vibroeng. **16**(2), 891–900 (2014)
11. Ding, H., Lim, C.W., Chen, L.Q.: Nonlinear vibration of a traveling belt with non-homogeneous boundaries. J. Sound Vib. **424**, 78–93 (2018)
12. Kim, M., Chung, J.: Dynamic analysis of a pulley-belt system with different pulley radii and support stiffness. J. Mech. Sci. Technol. **32**(12), 5597–5613 (2018). https://doi.org/10.1007/s12206-018-1106-8
13. Schnürer, D., Holl, H.J.: Transversal vibrations of a toothed belt in linear drives during operation. Proc. Appl. Math. Mech. (2021). https://doi.org/10.1002/pamm.202000026
14. Moon, J., Wickert, J.A.: Non-linear vibration of power transmission belts. J. Sound Vib. **200**(4), 419–431 (1997)
15. Łatas, W.: Active vibration suppression of axially moving string via distributed force. Vib. Phys. Syst. **31**, Article ID: 20202 (2020)
16. Lad, P., Kartik, V.: Stability transitions of an axially moving string subjected to a distributed follower force. P. Roy. Soc. A. **474**, Article ID: 20170779 (2018)
17. Pellicano, F., Vestroni, F.: Nonlinear dynamics and bifurcations of an axially moving. Beam. J. Vib. Acoust. **122**, 21–30 (2000)
18. Karli, D., Caji, M., Paunovi, S., Adhikari, S.: Periodic response of a nonlinear axially moving beam with a nonlinear energy sink and piezoelectric attachment. Int. J. Mech. Sci. **195**, Article ID: 106230 (2021)

19. Pham, P.-T., Hong, K.-S.: Dynamic models of axially moving systems: a review. Nonlinear Dyn. **100**(1), 315–349 (2020). https://doi.org/10.1007/s11071-020-05491-z
20. Zhu, H., Zhu, W.D., Fan, W.: Dynamic modeling, simulation and experiment of power transmission belt drives: a systematic review. J. Sound Vib. **491**, Article ID: 115759 (2021)
21. Bertini, L., Carmignani, L., Frendo, F.: Analytical model for the power losses in rubber V-belt continuously variable transmission (CVT). Mech. Mach. Theory. **78**, 289–306 (2014)
22. Chen, T.F., Sung, C.K.: Design considerations for improving transmission efficiency of rubber V-belt CVT. Int. J. Veh. Design **37**(4), 320–333 (2000)
23. Łatas, W., Kot, A.: Experimental and theoretical investigation of CVT rubber belt vibrations. Open Eng. **11**, 1–11 (2021)

Dynamic Model for Simulation of Phenomena Occurring in a Planetary Gear Operating in a Power Transmission System

Grzegorz Peruń(✉)

Silesian University of Technology, Katowice, Poland
grzegorz.perun@polsl.pl

Abstract. Vibration and noise accompanying the operation of gears are always unwanted phenomena with harmful effects on the environment. For this reason, it is necessary to work towards reducing these residual processes during gear operation. Such activities should be carried out already at the stage of gear design and take into account the assumed real operating conditions.

For economic reasons, machine downtime associated with maintenance should be as rare as possible and planned in advance after a certain period of operation. However, changes in operating conditions, as well as random factors, can cause failures to occur. Therefore, it is advisable to monitor the technical condition of the facilities and carry out their diagnostics in such a way as to reduce the risk of failure to a minimum, without significantly increasing operating costs.

The main goal of the presented research was to develop a nonlinear dynamic model of a planetary gear operating in a power transmission system, taking into account, among others an electric motor and a driven working machine. It is possible to use a dynamic model for the issues presented above, that is, for construction, optimization and diagnostics.

Keywords: Nonlinear dynamic model · Planetary gear · Diagnostics

1 Introduction

The operation of gears is accompanied by vibrations and the noise they cause. In the case of gears, these are always undesirable phenomena, having a negative effect on the environment, and thus also adversely affecting the assessment of the quality of work and competitiveness of a particular product on the market. To reduce these undesirable phenomena, appropriate solutions should be sought already at the design stage, guaranteeing effectiveness under the assumed operating conditions. Numerous studies and observations have shown that there are no solutions to achieve good properties in a wide range of operating parameters, but it is usually possible in a specific, assumed operating range. For this reason, especially in the case of gears that play an important role, costly gears, or gears whose possible failure could cause a threat to human life and health or cause large economic losses, it is essential to determine as accurately as possible the assumed, actual operating conditions before proceeding to design work. In

A. Puchalski et al. (Eds.): ICTD 2022, ACM 21, pp. 42–51, 2023.
https://doi.org/10.1007/978-3-031-31719-4_5

the case of already existing gearboxes, optimization are also possible, based on previous observations of the object's operation.

Vibration and noise are also factors that can inform about the technical condition of various technical objects, including gears. Changes in operating conditions, as well as random factors, can cause failures to occur. It is advisable to monitor the technical condition of the facilities and carry out their diagnostics in such a way as to reduce the risk of failure to a minimum, without significantly increasing operating costs. The monitoring systems used [4], which operate based on measurements of various quantities that also characterize residual phenomena, in effect make it possible to predict the occurrence of failures and to take appropriate corrective action in a reasonable time [19, 20]. This can prevent the emergence of a threat to human life and health or protect against economic losses. In some situations, the standing diagnostic procedures also make it possible to decide to delay a scheduled inspection. The possibility of correctly influencing the interval between inspections makes economic sense in particular. Machine downtime in many situations, especially in industrial conditions, should be as rare as possible and planned in advance after a certain period of operation [24].

Gears are devices that are widely used in the power transmission systems of various machines, so their construction and support of this process and diagnostics of simple gears, especially single-stage spur gears, with the use of dynamic models have already been dedicated to a lot of attention in the literature. The influence of various construction, technological, and operating factors on the possibility of reducing vibration and noise during the operation of simple one- and two-stage fixed-axis gears has been the subject of many scientific and research papers. It is quite different in the case of other types of gears, especially those whose design is definitely more complex. The group of such constructions includes planetary gears, which, due to their numerous advantages, are widely used in mechanical engineering and are used, among others, in the drive systems of motor vehicles (e.g. as retarders in the running wheels of trucks and buses, as well as speed reducers in starters), aircraft - mainly helicopters (e.g., as one/two output stages of the main reducer) and propeller-driven aircraft (in star-engine aircraft, where planetary reducers with spur or bevel wheels are encountered).

The degree of complexity of the construction of planetary gears and a large number of possible solutions make the task of designing and assessing the technical condition significantly more difficult. The main goal of the presented research was to develop a nonlinear dynamic model of a planetary gear operating in a power transmission system, taking into account, among others an electric motor and a driven working machine. The model allows studies to be carried out at variable speeds and loads, caused both by factors arising from the operation of the gear itself, as well as factors related to the operation of the other components of the power transmission system. It is possible to use a dynamic model for both of the issues presented in the introduction, that is, for construction, optimization and diagnostics. With its use, it is possible to define procedures for conducting measurements, signal processing, and analysis that will improve diagnostics of real gears.

2 A Brief Overview of Gears Modelling

The vibroacoustic activity of gears, classified as complex nonlinear mechanical vibration generators, has been the target of research by many scientific and research centers for years. In the literature one can find various approaches to modeling gears, often combining different methods, e.g. dynamic model with FEM model and/or analytical model. Detailed reviews of gear dynamics models with vibration emphasis can be found in review papers by Blankenship and Singh [5], Ozguven and Houser [17], Wang et al. [23] and Bartelmus [1, 2].

Many already historical models provide insight into the ways of modeling various phenomena, but they often made significant simplifications, required by the limitations of the nature of computational capabilities operating at the time of their development. Differences in the way of modeling often the same phenomena also resulted from the purpose of the models being developed, and the analysis of these different ways of working now allows the selection of the most appropriate approaches in the practically unlimited possibilities of conducting calculations.

The developed models can be classified according to various criteria. Taking into account the scope of modeling, one can distinguish between models of gearboxes isolated from the rest of the power transmission system components and models of gearboxes working in the power transmission system. The first group is characterized by a lack of consideration of the interactions of gears and the rest of the power train. The second group was most often characterized by significant simplifications in the description of the phenomena occurring in the gears.

Another criterion for dividing gear models can be to take into account the variation during the operation of gear pitch, which occurs in real gears. With significant pitch deviations, with a value greater than the static deflections of the teeth, there can be a temporary lack of contact between the mating teeth in the mesh zone, which means a change in the actual stiffness of the mesh. This phenomenon is not taken into account by models in which the gear stiffness is a constant value or varies as a function of time or angle of rotation. Advanced models that take into account the possibility of temporary loss of tooth contact include, for example, palisade models and models based on so-called apparent interference, widely presented, among others, in the publication [8].

For the study of planetary gears having manufacturing errors, to predict modulation sideband, a nonlinear time-varying dynamic model was proposed in the paper [11]. Another study to determine the effect of torque on the dynamic behaviour and system parameters of planetary gears, largely based on a measurement experiment, shown in the work [9], also presents extensively various models, including analytical and dynamic planetary gear. The other paper combining the modeling of wear and dynamic phenomena in planetary gears is [21]. A dynamic model of a planetary gearbox developed for dynamic simulation of eccentricity errors is also presented in [10]. Despite the limited number of degrees of freedom, complex effects of instantaneous rigid-body motions, couplings between errors were obtained, DOFs and gear geometry interactions, and modulated mesh stiffness functions. In paper [13] is studied a Ravigneaux planetary gear set system with an unloaded floating ring gear. The rigid dynamic model and three–dimensional dynamic load of the planet are presented. Obtaining vibration signals of a planetary gear in full efficiency and cracked tooth states is described in [14].

Another dynamic model of planetary gears, considering only one degree of freedom for each element - rotation around the wheel axis, among others, is presented in works [15, 16]. To study nonlinear dynamic phenomena in planetary gears, a two-dimensional FEM model and also a two-dimensional model with clustered parameters were used in the work [18]. The FEM model was used to determine the mesh stiffness and its variation, and boundary contacts and to model the loss of contact between teeth. The clustered parameter model reflected the wheels as masses with known moments of inertia and took into account the changing stiffness of the mesh represented as springs with nonlinear characteristics and stiffness depending on the contact condition of the teeth. This model is an extended version of the model developed by Lin and Parker [16], taking into account the phenomenon of loss of contact between teeth. The bearings were modeled as springs with linear characteristics. Frictional forces between the surfaces of the mating teeth were also taken into account.

The use of a dynamic model of planetary gears to simulate wheel tooth damage (pitting and tooth fracture) is presented in papers [6, 7]. The model of a single-stage planetary gear takes into account the linear stiffness of the bearings and the linear stiffness of the springs that represent the susceptibility of the mesh. Each component has three degrees of freedom: two displacements and one rotation. Damping was omitted from the model, but it is possible to account for it in parallel with the stiffness of the gearing and bearings. Local damage to the teeth was taken into account in the model by modifying the stiffness values accordingly. The study was aimed at determining the effectiveness of damage detection using Wigner-Ville analysis.

Dynamic models of complex planetary gears are also presented in the work [12]. The models consider torsional vibrations for 10 different gear configurations. In contrast, the paper [22] shows a dynamic model of a 2K-H type planetary gear. It takes into account the nonlinear, time-varying cooperative stiffness of the gears determined using three-dimensional FEM models.

3 Characteristics of the 2K-H Type Planetary Gear and Its Dynamic Model

Placing increasingly higher performance requirements on the design of gears results in the need to seek complex design solutions, and thus requires designers to solve an increasingly complex task. In addition, this task is made more difficult by the economic factor. Taking the above into account, the optimization of gear design that takes into account this growing number of often opposing criteria can increasingly take place only with the use of electronic calculation techniques, i.e. computer hardware with specialized software.

Due to its high popularity [3, 4], a 2K-H type planetary gear was selected for testing and analysis, the schematic diagram of which is shown in Fig. 1. In the design of this gear, the drive is fed to the sun wheel (1), and from it, power is fed to the output shaft through the planet wheels (p) and carrier frame (h). The ring wheel (2) remains stationary.

The characteristic frequencies of the gears, are related to the relative speeds of the sun wheel, expressed by the difference of the $n_1 - n_h$. The relative speed of the sun wheel

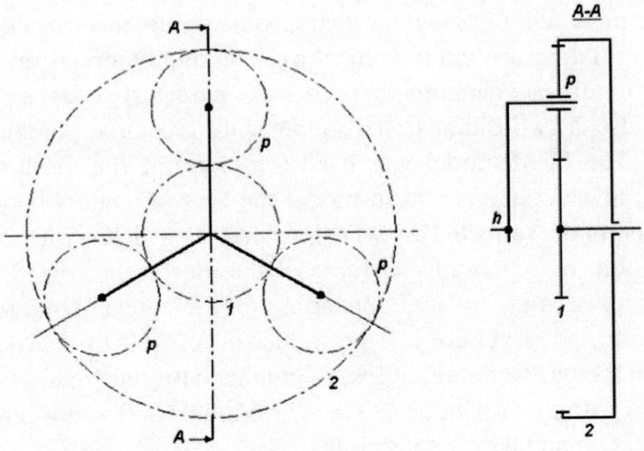

Fig. 1. Diagram of 2K-H type planetary gear: 1- sun gear, 2- ring gear, p- planet gears, h- carrier frame.

is expressed by the formula:

$$n_1 - n_h = n_A \frac{-i_0}{1 - i_0} \text{[rpm]} \tag{1}$$

where: n_1, n_h are the speeds of the sun wheel and carrier frame, respectively, n_A is the gear input speed in rpm, i_0 is the base gear ratio equal to z_2/z_1, which is the ratio of the number of teeth of the ring wheel to the number of teeth of the sun wheel.

The rotation frequency of the sun wheel can be determined from the relationship:

$$f_{01} = \frac{n_A}{60} \text{ [Hz]} \tag{2}$$

The frequency of rotation of the carrier frame, in turn, is determined by the formula:

$$f_{0h} = \frac{n_A}{60(1 - i_0)} \text{ [Hz]} \tag{3}$$

Based on the analysis of the literature on dynamic models of gears, including planetary gears, and taking into account current trends in modeling, it was decided to develop a dynamic model of a planetary gear operating in a power transmission system.

With the concept of modeling the entire power transmission system is possible to take into account both internal interactions, the source of which is the gear itself, as well as external interactions coming from the engine and the driven device, and the interactions that occur between them. It was also assumed that the model should be characterized by high accuracy in describing the phenomena occurring in the mesh and allow simulating the operation of the gearbox under different operating conditions - mainly at different loads and speeds. Due to one of the adopted objectives of the research, the model from the beginning should be characterized by the ability to simulate local damage to bearings and gears teeth.

The extended full model of the transmission system consists of a planetary gear, an asynchronous electric motor, a driven machine (with the known moment of inertia and torque characteristics), shafts, bearings, and couplings connecting selected elements. A simplified schematic of the model is shown in Fig. 2. The wheels of the gear, the working machine, and the couplings are treated in the model as rigid bodies with known moments of inertia, connected by weightless elements modeling elasticity and damping. The masses of the other gear elements were reduced to masses focused on the centers of the bearings.

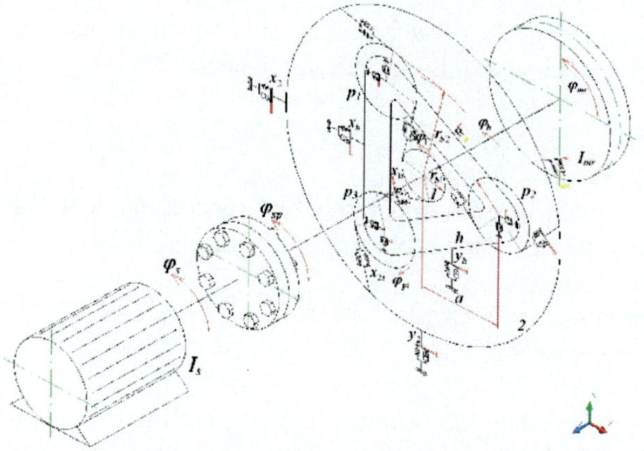

Fig. 2. A dynamic model of 2K-H type planetary gear in the power transmission system.

The excitation of mass vibration components is the result of the complex interactions of both internal sources of vibration, resulting from the gear's performance of the objective function, which includes varying mesh stiffness, damping in the mesh and mesh deviations, and external sources, resulting from changes in shaft speeds and torques.

The dynamic model of the propulsion system, as a result of the simplifications used, was described using the following generalized coordinates q_i:

- angular displacement:
  - rotor around the axis z: φ_s,
  - couplings around the axis z: φ_{sp},
  - sun wheel and ring gear relative to the axis z: φ_1, φ_2,
  - planetary gears around their axes φ_{pi},
  - carrier frame around the axis z: φ_h,
- radial displacements:
  - sun and ring gears in the direction of the axis $\boldsymbol{x}$: x_1, x_h,
  - sun and ring gears in the direction of the axis $\boldsymbol{y}$: y_1, y_h,

- displacements in the bearings of planetary wheels: x_{pi}.

The most important features of the model and parameters of the power transmission system and the gearbox operating in it are as follows:

- basic design features of gear:
  - number of circulating wheels,
  - various diameters of the wheels,
  - phase shifts of the gears between the planetary wheels and the sun wheel and between the planetary wheels and the ring wheel;
- mesh parameters:
  - deviations of pitch, profile and tooth line (according to PN-ISO 1328:2000), declared separately for each tooth in each wheel,
  - clearances in the gears,
  - modification of the tooth profile,
  - modification of the tooth line;
- variable mesh stiffness along the contact line - the well-known method developed by Niemann and Baethge was used, its details are presented among others in the work [8];
- damping in the gear mesh;
- nonlinear stiffness of all bearings;
- vibration damping in the gear bearings (including the planetary wheels).

A prerequisite for obtaining the assumed functionality of the model was the development of an algorithm for determining the numbers of currently cooperating teeth in each gear mesh. Taking into account the correct numbers of the teeth during calculations makes it possible to determine the total deviations of each pair of teeth located in a given computational step in the gear mesh, based on the previously declared error values, and makes it possible to simulate various local damage to the gears.

The dynamic model of the planetary gear drive system described above was realized in the Simulink simulation environment. Due to numerous optimization procedures, this environment allowed acceptable calculation times. In Matlab, on the other hand, the basic calculations of the data necessary for the simulation are realized. Simulation results are saved directly to files in Matlab format, which allows further processing without the use of data import procedures.

There are three main steps in conducting research using the software developed:

- data declaration and basic pre-simulation calculations,
- simulation of power transmission system operation at the assumed time,
- analysis of the obtained results, preceded by possible resampling in the case of choosing a variable calculation step and/or decimation of output signals.

4 Possibilities of Applying the Model

Thanks to the use of the model, it is possible to obtain, among other things, real values of displacements, velocities, and accelerations of individual elements from the whole or selected time interval of the performed calculations, time courses of forces in each mesh, courses of forces transmitted by individual pairs of teeth (also in the form of spatial diagrams) as well as forces in bearings. The determined values are consistent with the results of experiments performed on the test stand, which allowed, at an earlier stage of the work, to determine all the parameters of the selected gear and to tune the model.

Examples of force time-courses in the mesh of the sun wheel with one of the planetary wheel and in the mesh of the ring wheel with the planetary wheel, under the idealistic assumption of no manufacturing deviations, are shown in Fig. 3. Number of teeth $z_1 = 27$, $z_p = 38$, $z_2 = -102$, straight mesh, module m = 1,5 [mm], nominal contact angle $\alpha_0 = 20°$, pitch diameter $d_1 = 40{,}5$ [mm], $d_p = 57$ [mm], $d_2 = -153$ [mm], mesh width $b = 28$ [mm]. Approximate masses and moments of inertia of gear wheels: $m_1 = 150$ [g], $m_p = 200$ [g], $m_2 = 960$ [g], $I_1 = 40$ [kg mm^2], $I_p = 130$ [kg mm^2], $I_2 = 6000$ [kg mm^2]. Average stiffness of the sun wheel and the planetary wheel mesh 12.9 [N/mm µm], the planetary wheel – ring gear mesh 15.3 [N/mm µm].

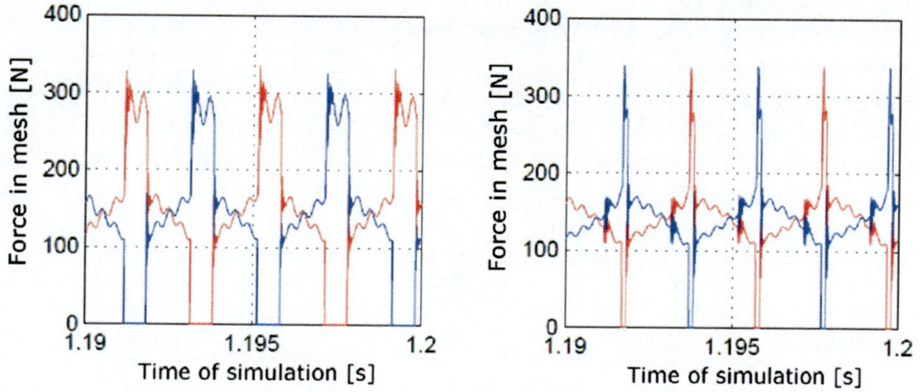

Fig. 3. Examples of forces in the mesh: a) a sun gear with one of the planetary gears, b) a ring gear with one of the planetary gears; input shaft speed $n_1 \approx 1457$ rpm; wheels without teeth deviation.

Based on the force time-courses, it is possible to determine the influence of specific factors on the vibration performance of gears. In addition, these courses can be used for testing with FEM models of various gear body designs. The courses of other quantities can be used both for design optimization and for the development of diagnostic algorithms sensitive to the appearance of malfunctions. This is made possible by the introduced possibility of simulating local failures of bearings and wheel teeth in any of their configurations. There is also a simplified possibility to simulate a gear characterized by wear on the working surfaces of the teeth. However, the presentation of sample results of gearbox operation with simulated damage is beyond the volume of this paper.

5 Summary

The determined parameters of gear geometry, including wheels, shafts and bearings were entered into the model and allowed to state the high correspondence of the obtained results of numerical tests with the results of tests carried out on the real object. The similarity of the results of numerical tests and experimental studies makes it possible, by means of simulation calculations, to determine the phenomena occurring in the gear mesh under different operating conditions.

For the above reason, the developed dynamic model is currently being used in a wide range of research. The original data describing the power transmission system with planetary gear are modified accordingly, which makes it possible to determine the influence of selected design and operating factors on the operation of the analyzed gear. In each case, the values of the forces in the wheels and bearings are calculated, which makes among others possible to determine the overloading of the teeth beyond the nominal value, as well as the unevenness of the load distribution on the individual planetary wheels. The courses and values of forces in meshes and forces in the bearing nodes can be a good indicator of the degree of optimization of the gear design.

Other research realized with the dynamic model confirms that the minimization of vibroacoustic activity in gearboxes should be carried out by appropriate reduction of forces in the gears, by means of proper selection of gearbox design features. However, it should be remembered that the effectiveness of minimizing vibroacoustic activity is influenced by bearing nodes and the gearbox body.

Studies have also shown that the occurrence of damage in the early stages of development, is even more difficult to diagnose than in simpler constructions of gear and often proves impossible with vibroacoustic methods. It was found, that due to the different symptoms of damages and the possibility of simultaneous occurrence of several damages at different stages of development, the detection of malfunctions is facilitated by comparing the results of signal analyses with the results of those analyses carried out using signals recorded when the object was in a fully operational state.

References

1. Bartelmus, W.: Mathematical modelling and computer simulations as an aid to gearbox diagnostics. Mech. Syst. Signal Process. **15**(5), 855–871 (2001)
2. Bartelmus, W.: Supporting diagnostic inference by mathematical modelling from one-stage to planetary gearbox systems. Diagnostyka **30** (2004)
3. Bartelmus, W., Zimroz, R.: Identification of operating conditions for the diagnosis of a planetary gear for bucket wheel drive [In Polish: Identyfikacja warunków eksploatacyjnych na potrzeby diagnostyki przekładni planetarnej do napędu koła czerpakowego]. Diagnostyka **1**(37) (2006)
4. Batko, W., Korbiel, T.: Diagnostic monitoring system for planetary gears operating under variable load conditions [In Polish: System monitoringu diagnostycznego przekładni planetarnej pracującej w warunkach zmiennego obciążenia]. Diagnostyka **1**, 77–82 (2006)
5. Blankenship, G.W., Singh, R.: A comparative study of selected gear mesh interface dynamic models. In: ASME International Power Transmission and Gearing Conference (1992)
6. Chaari, F., Fakhfakh, T., Haddar, M.: Dynamic analysis of a planetary gear failure caused by tooth pitting and cracking. J. Fail. Anal. Prev. **6**(2) (2006)

7. Chaari, F., Fakhfakh, T., Haddar, M.: Numerical simulation of planetary gear failure detection by Wigner-Ville method. In: Second World Congress on Engineering Asset Management and the 4th International Conference on Condition Monitoring. Harrogate, UK (2007)
8. Dąbrowski, Z., Radkowski, S., Wilk, A.: Gear dynamics. In: Testing and Simulation in Operationally Oriented Design [In Polish: Dynamika przekładni zębatych. Badania i symulacja w projektowaniu eksploatacyjnie zorientowanym]. Wydawnictwo i Zakład Poligrafii Instytutu Technologii Eksploatacji, Warszawa – Katowice – Radom (2000)
9. Ericson, T.M., Parker, R.G.: Experimental measurement of the effects of torque on the dynamic behaviour and system parameters of planetary gears. Mech. Mach. Theory **74**, 370–389 (2014)
10. Gu, X., Velex, P.: On the dynamic simulation of eccentricity errors in planetary gears. Mech. Mach. Theory **61**, 14–29 (2013)
11. Inalpolat, M., Kahraman, A.: A dynamic model to predict modulation sidebands of a planetary gear set having manufacturing errors. J. Sound Vib. **329**, 371–393 (2010)
12. Kahraman, A.: Free torsional vibration characteristics of compound planetary gears sets. Mech. Mach. Theory **36** (2001)
13. Lai, J., et al.: Dynamic modeling and analysis of Ravigneaux planetary gear set with unloaded floating ring gear. Mech. Mach. Theory **170** (2022)
14. Liang, X., Zuo, M.J., Hoseini, M.R.: Vibration signal modeling of a planetary gear set for tooth crack detection. Eng. Fail. Anal. **48**, 185–200 (2015)
15. Lin, J., Parker, R. G.: Planetary gear parametric instability caused by mesh stiffness variation. Journal of Sound and Vibration 249(1) (2002)
16. Lin, J., Parker, R.G.: Sensitivity of planetary gear natural frequencies and vibration modes to model parameters. J. Sound Vibr. **228**(1) (1999)
17. Ozguven, H.N., Houser, D.R.: Mathematical models used in gear dynamics - a review. J. Sound Vib. **121**, 383–411 (1988)
18. Parker, R.G., Lin, J.: Modeling, modal properties, and mesh stiffness variation instabilities of planetary gears. NASA Final Contractor Report (2001)
19. Pawlik, P.: The use of the acoustic signal to diagnose machines operated under variable load. Arch. Acoust. **45**(2) (2020)
20. Pawlik, P., et al.: Vibroacoustic study of powertrains operated in changing conditions by means of order tracking analysis, Eksploatacja i Niezawodnosc – Maintenance Reliab. **18**(4), 606–612 (2016)
21. Shen, Z., Qiao, B., Yang, L., Luo,W., Yang, Z., Chen, X.: Fault mechanism and dynamic modeling of planetary gear with gear wear. Mech. Mach. Theory **155** (2021)
22. Sun, T., Hu, H. Y.: Nonlinear dynamics of a planetary gear system with multiple clearances. Mechanism and Machine Theory 38 (2003)
23. Wang, J., Li, R., Peng, X.: Survey of nonlinear vibration of gear transmission systems. Appl. Mech. Rev. **56**(3), 309–329 (2003)
24. Živković, P., Milutinović, M., Tica, M., Trifković, S., Čamagić, I.: Reliability evaluation of transmission planetary gears "bottom-up" approach. Eksploatacja i Niezawodnosc – Maintenance Reliab. **25**(1) (2023)

Acoustic Emission Monitoring for Wire Drawing Process

Enrique Caso(✉), Pablo García-Fernández, Alfonso Fernández-del-Rincón, Ana De-Juan-De-Luna, Alberto Díez-Ibarbia, and Javier Sánchez-Espiga

Ingeniería Mecánica, Universidad de Cantabria, Avenida de los Castros S/N, Santander, Spain
casoe@unican.es

Abstract. Wire drawing is a process consisting on the reduction in the section of a wire by means of its elongation. The smaller cross-section is obtained pulling the wire at room temperature through the die. Plastic deformation is complemented by friction between wire and die surfaces. Modifications of the operating conditions and the parameters related with the surface contact and lubrication favour material damage and die deterioration. Wire drawing is subject of incidences and failures that lead to a waste of time and resources. The failures can be undetected with current monitoring technologies, therefore, the interest for the evaluation of innovative sensors. However, the acoustic emission is regarded as a promising tool for process control and defect diagnosis due to the physical phenomena within the wire-die interaction. In this work, an analysis of experimental data is developed using acoustic emission signals obtained in the monitoring of an industrial steel wire drawing process. Slight modifications in the main operational parameters, as drawing speed, temperature or lubricant characteristics, are correlated with acoustic emission signals. Therefore, the signal response is validated for the process control in consequence of its sensitivity to deviations associated with abnormal conditions or damage appearance.

Keywords: Wire drawing · Acoustic emission · Condition monitoring

1 Introduction

Some research has been carried out to develop tools for wire drawing monitoring based on different sensor technologies. The technologies evaluated include, among others, vibration analysis [1], digital image processing to identify surface modifications by means of surface reflection variation [2], thermographic image processing of wire temperature [3] or a combination of this information [4]. Also by measuring tensile strength [5], lubricant layer electrical resistance [6,7] and eddy currents to detect surface defects on the wire [8]. On the other hand, acoustic emission (AE) waves are generated in physical mechanisms present in

A. Puchalski et al. (Eds.): ICTD 2022, ACM 21, pp. 52–62, 2023.
https://doi.org/10.1007/978-3-031-31719-4_6

the wire-die interaction. Although the literature is lacking in this area [9,10], there are proposals for the application of AE tools in the wire drawing industry [11,12]. The preliminary nature of some of these studies and the lack of research on this subject make necessary further analysis to establish an adequate AE monitoring of the wire drawing process, successfully used in other fields [13,14].

The drawing involves plastic deformation of the wire, due to volumetric and shear strain, as it passes through the die. The die is the main component, made of a high hardness material with a hole smaller than the incoming wire cross-section. AE is generated during plastic deformation due to the dislocations movement inside the material subjected to stress. Furthermore, drawing involves mechanisms of an elastic nature in the interaction between the wire and die surfaces, hence capable of emitting AE when the stored energy in the asperity contact is suddenly released. Two aspects are fundamental in the process control as well as in AE generation:

- The type of lubricant, whose lubrication features reduce friction and abrasive phenomena on contact.
- The temperature, which rises due to the heat generated during plastic deformation and displacement under friction varying the process conditions, which, in turn, can modify the properties of the lubricant.

1.1 Influence of the Lubricant on the Drawing and the Generated AE

The industrial processes imposes high drawing speeds to maximise the wire production, constrained by the fact that an excessive speed increases the risk of incidents. Therefore contact is considered to occur in a boundary lubrication regime.

Boundary lubrication implies that both lubricant and surface properties influence the contact. In this case, friction is controlled by the chemical properties of the lubricant rather than its viscosity. The lubricant reacts with the wire surface to create a softer material than die or the wire itself. As a result, this softer surface deforms more easily, protecting the die and wire from wear [16]. The lubricant changes from powder to a plastic state. This change in properties, that allows the powder-type-lubricant to become a protective layer, takes place once it reaches the working zone in the die intake cone.

The main lubricant component for wire drawing soaps is usually a fatty acid soap, which is combined with other additives that modify its properties. Dry soap type lubricants can be classified according to their solubility in water, establishing two categories:

1. Sodium soap (soluble): fatty acid + caustic soda → sodium soap + H_2O.
2. Calcium soap (insoluble): fatty acid + lime → calcium soap + H_2O

Calcium soap is recommended for the early stages, while sodium soap is used in the later stages where the wire reaches higher speeds and stresses [17]. Sodium soaps are more fluid than calcium soaps [16] and therefore favour conditions closer to hydrodynamic lubrication than calcium soap. Another consequence of the soap properties is that the shear stresses in the wire are lower using a sodium soap than a calcium soap [17].

1.2 Temperature Effect on the Process and the AE

Wire drawing is a cold process, therefore an excessive increase in temperature has an adverse effect, contributing to wire breakage and decreasing wire ductility [15, 18–20]. A problem involving a malfunction of the die cooling system implies an overheating of its active surface.

A refrigeration failure imposes a new thermal equilibrium. The temperature rise is transferred to the lubricant and the entire block where the die is located. As a consequence, a cooling problem has several effects that affect the process and can be reflected in the AE.

Firstly, the temperature increment modifies the lubricant performance. The lubricant is affected, degrading under extreme conditions, reacting with the wire surface or changing its viscosity and wire wettability.

On the other hand, an increase in temperature reduces the stress flow in the wire and the required drawing force. This affects the AE due to plastic deformation owing to its dependency on the stress-strain curve point describing the material behaviour [22, 23].

Additionally, the overheating has an effect on the die, generating its thermal expansion and modifying the nominal diameter. Thermal effects have an effect on the resulting product quality [21] and reducing the hardness of die surface, consequently, favouring its degradation [25]. To cool the die is effective in reducing its abrasion, resulting in improved surface conditions that provide a better final wire quality [26].

2 Experimental Procedure

The experiments were developed in a wire drawing factory under working conditions, without significant modifications to the standard procedure. Therefore, that supposes a limitation in the control of the operational variables, however, the monitoring results are more significant. The last stage of a wire drawing line was monitored for the tests (Fig. 1), as it concentrates most of the incidents and breakages due to the smallest wire section.

The AE signals were acquired using a *VS150-RIC* resonant type AE sensor with a peak frequency at 150 kHz positioned using a magnetic holder (Fig. 1). The sampling rate used was 1 MS/s, recording periodically signals of 1 s duration. In addition to the AE signals, wire speed (input and output) and wire temperature (input and output) were recorded. The die temperature is referred to the external surface of the die-block containing the die.

Two tests were defined to assess the AE sensitivity to the drawing process:

1. Lubricant type comparison test: The test is performed during the continuous drawing of one wire (no change of the wire material or modifications in the

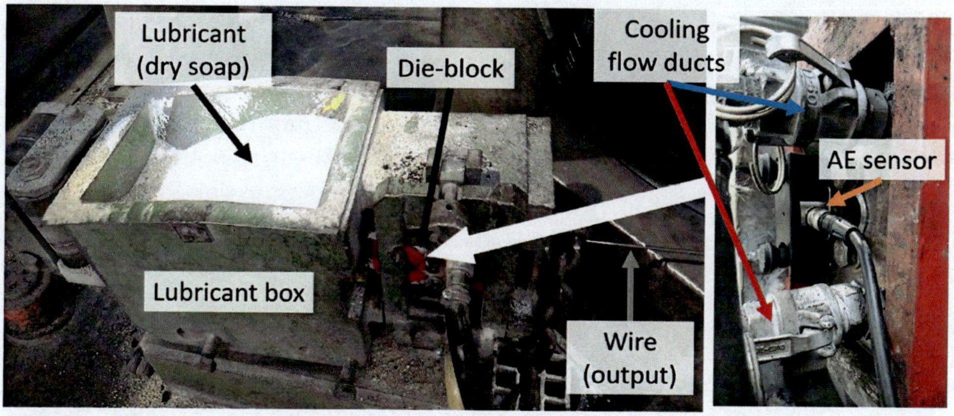

Fig. 1. Last stage of the drawing line and AE sensor position.

drawing line components). It is divided into three stages in order to analyse the effect of different lubricants on the AE signals. An identical operation consisting on drawing at three stationary speeds is repeated (Fig. 2). The only variation resides in the employed lubricant type. The third step repeats the first one, verifying that results are due to the lubricant change effect.

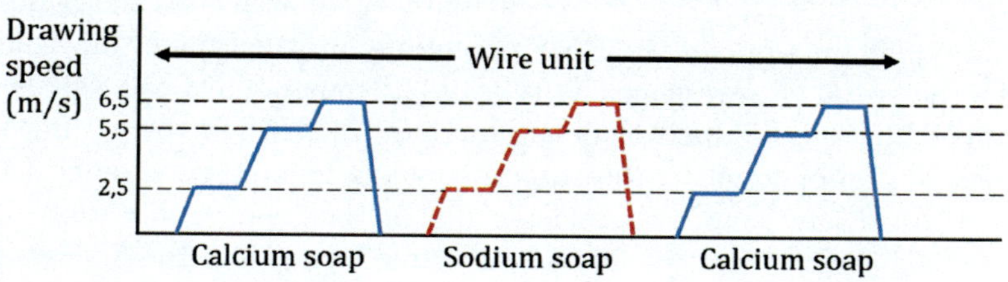

Fig. 2. Test procedure for lubricant comparison test.

2. Die overheating test: A cut-off of the cooling flow is performed during drawing, resulting in an increase in the die temperature. This simulated incidence is initiated after a warm-up and stabilisation period after achieving stationary conditions at a speed of 6.5 m/s. The whole test is performed within the same continuous wire and invariant lubricant type. To monitor the evolution, AE signals are acquired periodically to interpret the response to the changes due to overheating.

3 Results

3.1 Effect of Lubricant Type on AE

The AE signals amplitude varies according to the used type of lubricant, considering same operating conditions. In this case, the emissions generated using calcium lubricant show lower amplitude than with sodium lubricant at the same drawing speed. The AE amplitude is affected by the drawing speed. AE amplitude increases when the speed is increased, independently of the lubricant used. Therefore, the signals at 6.5 m/s have a higher RMS than those at 5.5 m/s and the latter have a higher RMS than those acquired at 2.5 m/s for both lubricants.

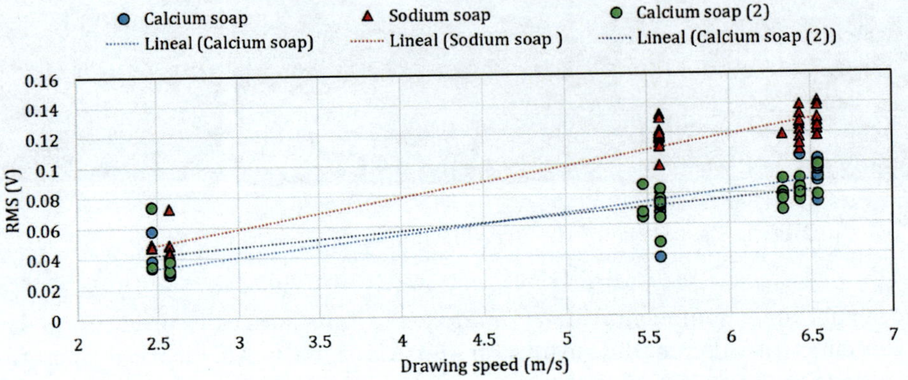

Fig. 3. AE RMS values at different speed and lubricant type.

Figure 3 shows the RMS of the AE signals, calculated for each acquisition. It shows the different slope of the RMS depending on the type of lubricant with respect to the speed. Expressing the relative increment between values at the different speeds, it is obtained that the increase in RMS with calcium soap is approximately proportional to the square root of the speed increment. On the contrary, with sodium soap, this relationship in the relative variation between RMS and speed is close to be proportional. Same trend than the first stage with calcium soap is repeated in the third one (repetition drawing with calcium soap), corroborating the different AE response depending on the type of lubricant that involves different contact behaviour between surfaces.

Figure 4 shows spectra of the AE signals acquired during the test, where three regions are observed for each lubricant due to the different speed. The spectra from the signals recorded under sodium soap lubrication display higher peak values and a larger spectral area over the entire frequency range, determined by the sensor frequency response. Comparing the spectra of first calcium stage (Fig. 4a) and second calcium stage (Fig. 4c), a similarity in the morphology of the spectrum is shown, with minor variations between signals at 5.5 and 6.5 m/s.

Variations in the drawing control parameters, as speed and temperature, have an interconnected effect in the AE response. Therefore, changes in the AE signals must be compared in respect to the operating variables. Otherwise, a variation of another parameter could be the cause of the effects associated with the type of lubricant. The speed is controlled to be the same in all test stages and under stationary conditions. The wire and die temperature is likely to be affected in the process. However, negligible deviations have been observed in speed or wire and die temperatures between similar test conditions. In addition to this, the wire properties (roughness, surface treatment, casting composition, cross-section, ...) can affect the AE, nevertheless the test has been performed in one continuous wire. Thus, the results are valid within the same wire, maintaining the same observed AE response depending on the soap type.

However, when the test with same calcium and sodium soap was repeated, drawing a wire of different characteristics, the obtained results are contradictory. In the second test, an inverse signal amplitude response has been obtained in respect to the type of lubricant: calcium soap presents signals of greater amplitude than sodium soap. By using the same control speeds on both wires, the same contact conditions were not exactly being imposed on the two tests. The wires differed in the type of casting material they are made of and the dies used were different and their degree of wear cannot be assumed to be equivalent.

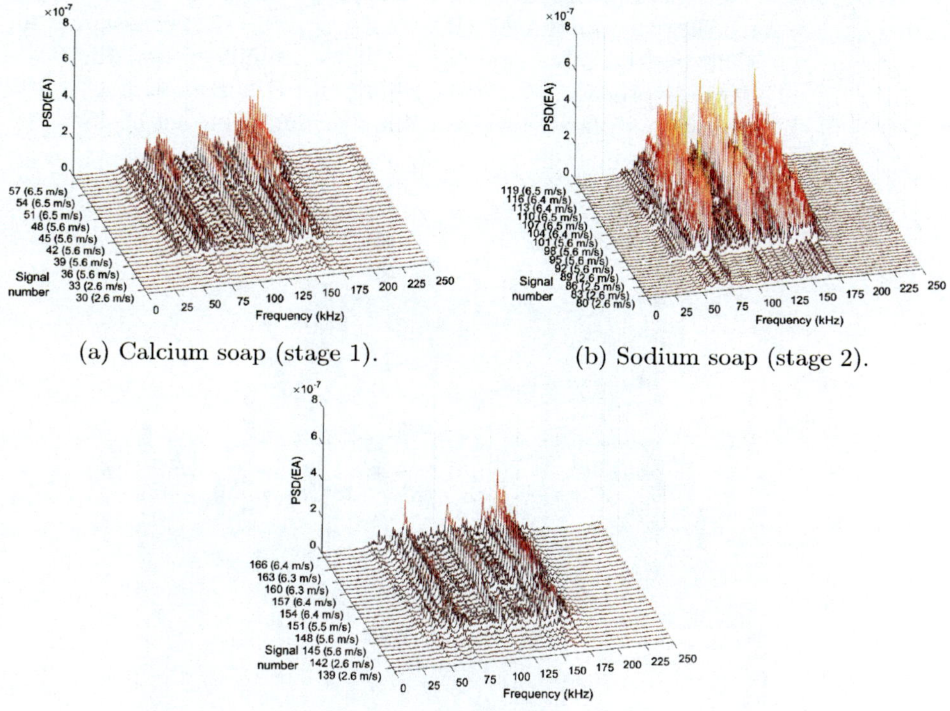

(a) Calcium soap (stage 1).

(b) Sodium soap (stage 2).

(c) Calcium soap (stage 3).

Fig. 4. AE spectrum for the two lubricant types at three speed levels.

The different roughness, material properties and the reactivity of the lubricant are reflected in the AE signals due to the sensitivity to slight variations in the process. It seems that a consistent interpretation can only be achieved within the same wire, or introducing more parameters to the actual monitoring system to quantify those aspects. Nevertheless, the AE analysis presents a different amplitude with a different chemical composition of the lubricant. Therefore, the analysis carried out shows the sensitivity and dependence of the generated AE on the parameters affecting the drawing process.

3.2 Die Overheating Effect on AE

The effect of an overheating due to the cut-off of the cooling in the die was evaluated, simulating an occurrence with these characteristics. The die heating affects increasing the temperature at the contact interface, modifying the behaviour of the wire, lubricant and die interaction.

Figure 5 shows the evolution of the AE RMS of the periodically acquired signals with two minutes interval. It starts with a warm-up period until process becomes stationary at 6.5 m/s, avoiding transitory effects in the AE. After that, the refrigerant is cut off. Two minutes after shutdown the water flow (signal number 30) a sharp decrease in the AE RMS is observed. Between signal 30, recorded just after closing the valve, and signal 32, two minutes later, the RMS is reduced by 14%. On the other hand, the die temperature increases from 26.8°C to 29.2°C in the same period, far below from the warning value set at 45°C.

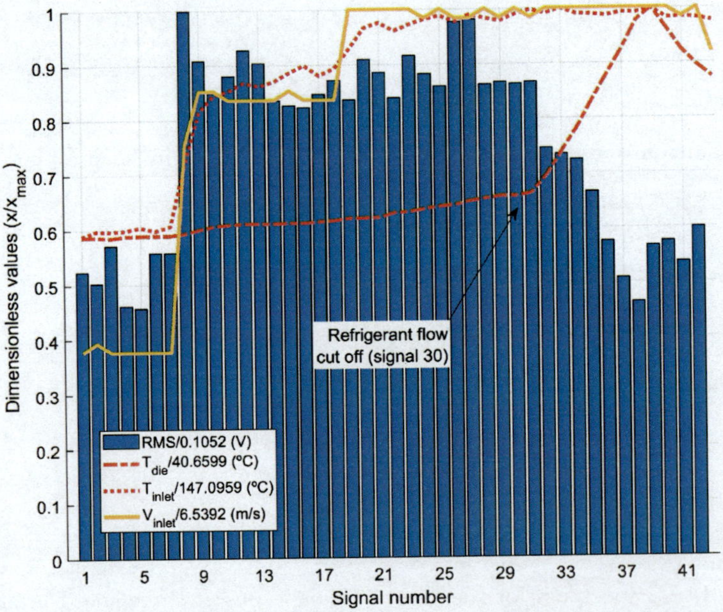

Fig. 5. AE RMS values versus temperature and speed (dimensionless).

The AE RMS continues to decrease progressively at same time that the die temperature increases. Between signals just before the cut-off and the last signal before opening (at 40.6°C), the RMS is reduced by 45%, without changing any control parameters of the operation. On the other hand, the exit wire temperature after the die remains virtually unvarying despite the cooling cut-off.

Figure 6 displays the test evolution of the AE frequency spectrum. AE spectrum follows the RMS response due to the overheating effect, achieving an overall reduction of its amplitude. However, the decrease of peak values in the spectrum is more pronounced in the low frequency band between 25-125 kHz. AE in this band is rapidly reduced, displaying a characteristic pattern that returns to normal spectrum levels when the cooling is reactivated. The spectral content at the AE sensor characteristic frequency (resonant around 150 kHz) is reduced but remains distinguishable in spite of the overheating.

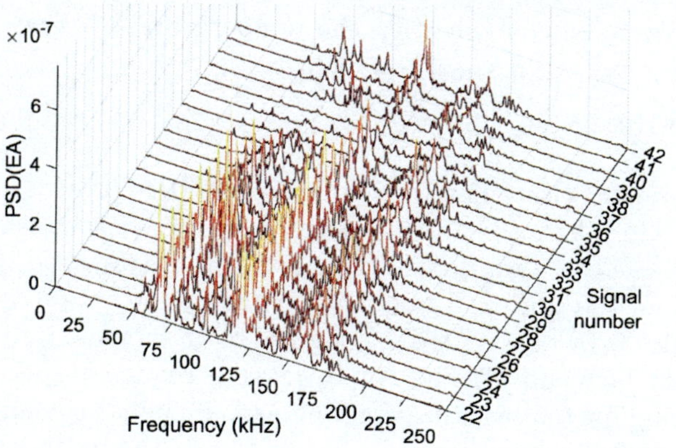

Fig. 6. Evolution of the AE spectra after cooling cut-off.

Repeating the test with a different characteristics wire, the same behaviour observed in the AE due to the cooling cut-off is reproduced. A progressive reduction of AE amplitude is observed, independently of the different spectral pattern in the wire, distinct from the previous one.

The interpretation of these results is based on two aspects related to the higher die temperature. This higher value means that the heat flow from the friction zone to the block is lower due to the lower temperature gradient. However, no significant increase in the temperature of the wire was observed at the output after the section reduction. Therefore, the lubricant would be mainly affected by an increase in temperature in the contact area, and the lubrication regime and chemical reactivity with phosphates in the wire surface would be modified.

The continuous AE emission would be partly caused by the effect of the lubricant. As no single AE events were observed, which could be associated with micro-contacts or galling, a higher temperature could be favouring a lubrication that resembles hydrodynamic. The lower AE emissivity would be related with the interface between wire and die. This result points to the importance of achieving an optimal working temperature, which is not controlled in the current process. However, increasing temperature and reducing AE does not have to be an improvement. A phosphate pre-coating provides a better wire surface to drag the lubricant and improves its properties in terms of viscosity, hydrodynamic lubrication and boundary lubrication. However, this coating is easily reacted, degrading wire surface after the wire exits the die due to the overheating.

Another interpretation resides in the effect on the wire. A softer material at the contact interface reduce its emissions due to the elastic nature friction mechanism between asperities, in spite of the boundary lubrication regime. The higher temperature due to the overheating affects the stress-strain properties of the tensile wire. During a tensile test the AE amplitude decreases after the yield point, obtaining minimum values before the crack initiation that precedes the breakage [23]. The AE would have a higher importance of plastic deformation compared to a frictional contact when the temperature is lower.

4 Conclusions

The use of a lubricant with different chemical properties provides an appreciable response in the AE. The lubricant interferes with the contact between wire and die active surface. Therefore, depending on the lubricant type, the AE signals show differences in the time and frequency domain.

AE signals also provide a sensitive response to an increase in die temperature due to refrigerant flow cut-off. The AE RMS and frequency spectrum decreases with the increasing die temperature. This effect is linked to a change in the working point of the lubricant and material properties of the involved components.

The obtained results show the suitability of the AE comparing lubricants in order to establish the optimum type, operating speed and working temperature. However, the conclusions cannot be extrapolated for all cases due to the opposite results varying lubricant type in different wires. Therefore, the presented analysis should be mainly interpreted in terms of relative variation. To stablish global conclusions requires to consider other parameters according to the wire an die properties that are not taken into account in this manuscript.

References

1. Pejryd, L., Larsson, J., Olsson, M.: Process monitoring of wire drawing using vibration sensoring. CIRP J. Manuf. Sci. Technol. **18**, 65–74 (2017). https://doi.org/10.1016/j.cirpj.2016.09.006
2. Larsson, J., Jansson, A., Pejryd, L.: Process monitoring of the wire drawing process using a web camera based vision system. J. Mater. Process. Technol. **249**(April), 512–521 (2017). https://doi.org/10.1016/j.jmatprotec.2017.06.020
3. Larsson, J., Jansson, A., Karlsson, P.: Monitoring and evaluation of the wire drawing process using thermal imaging. Int. J. Adv. Manuf. Technol. **101**(5–8), 2121–2134 (2019). https://doi.org/10.1007/s00170-018-3021-7
4. Larsson, J., Larsson, A., Pejryd, L.: Wire 4.0.pdf. Wire J. Int., 58–63 (2020). https://www.wirenet.org

5. Stopa, M.M., Cardoso Filho, B.J.: Online torque and drawing force estimation in wire drawing process from electric motor variables. IEEE Trans. Ind. Appl. **44**(3), 915–922 (2008). https://doi.org/10.1109/TIA.2008.921449
6. Nilsson, B., Stenlund, B.: Detection of lubrication failures in wire drawing. Wire Ind. **611**(51), 855–858 (1984)
7. Holm T, N.B., Karlstrom, K.E., Philipson, A.: Lubrication failures in wire drawing. Wire Ind. **611**(51), 242–245 (1985)
8. Foerster Group. https://www.foerstergroup.de/en/deu/
9. Masaki, S., Tabata, T., Konishi, K.: Evaluation of lubrication in wire drawing using acoustic emission method. J. Jpn Soc. Techn. Plast. **295**(25), 835–841 (1985)
10. Choi, J.D., Lee, M.Y., Park, W.G.: Acoustic emission monitoring fine wire drawing process. J. Korean Soc. Mach. Tool Eng. **3**, 43–50 (1996)
11. QASS. https://www.qass.net/wp/case-studies/wire_drawing_monitoring/
12. Pease, N.C.: Flaw detection in wire drawing," US4538463A (1983)
13. Caso, E., Fernandez-del-Rincon, A., Garcia, P., Iglesias, M., Viadero, F.: Monitoring of misalignment in low speed geared shafts with acoustic emission sensors. Appl. Acoust. **159**, 107092 (2020). https://doi.org/10.1016/j.apacoust.2019.107092
14. Caso, E., Fernandez-del-Rincon, A., Garcia, P., Diez-Ibarbia, A., Sanchez-Espiga, J.: An experimental study of acoustic emissions from active surface degradation in planetary gears. MSSP **189**, 110090 (2023). https://doi.org/10.1016/j.ymssp.2022.110090
15. Byon, S.M., Lee, S.J., Lee, D.W., Lee, Y.H., Lee, Y.: Effect of coating material and lubricant on forming force and surface defects in wire drawing process. Trans. Nonferrous Met. Soc. China (English Ed.) **21**(SUPPL. 1). s104–s110 (2011). https://doi.org/10.1016/S1003-6326(11)61071-6
16. Montmitonnet, P., Brison, M., Delamare, F.: Metallic soap lubrication in wire-drawing: A thermopseudoplastic model. Wear **77**(3), 315–328 (1982). https://doi.org/10.1016/0043-1648(82)90057-6
17. Felder, E., Levrau, C., Mantel, M., Truong Dinh, N.G.: Identification of the work of plastic deformation and the friction shear stress in wire drawing. Wear **286–287**, 27–34 (2012). https://doi.org/10.1016/j.wear.2011.05.029
18. Nakamura, Y., Kawakami, H., Fujita, T., Yamada, Y.: The new cold wire drawing method. Bull. Japan Inst. Met. **17**(2), 140–141 (1978). https://doi.org/10.2320/materia1962.17.140
19. Lee, S.K., Ko, D.C., Kim, B.M.: Pass schedule of wire drawing process to prevent delamination for high strength steel cord wire. Mater. Des. **30**(8), 2919–2927 (2009). https://doi.org/10.1016/j.matdes.2009.01.007
20. Raemdonck, W.V., Lefever, I., D'Haene, U.: Torsion tests as a tool for high strength wire evaluation. Wire J. Int., 68–75 (1994)
21. Moon, C., Kim, N.: Analysis of wire-drawing process with friction and thermal conditions obtained by inverse engineering. J. Mech. Sci. Technol. **26**(9), 2903–2911 (2012). https://doi.org/10.1007/s12206-012-0711-1
22. Schnabel, S. Golling, S., Marklund, P., Larsson, R.: The influence of contact time and event frequency on acoustic emission signals. Proc. Inst. Mech. Eng. Part J. J. Eng. Tribol., **231**(10), 1341–1349 (2017). https://doi.org/10.1177/1350650117693176
23. Akbari, M., Ahmadi, M.: The application of acoustic emission technique to plastic deformation of low carbon steel. Phys. Procedia **3**(1), 795–801 (2010). https://doi.org/10.1016/j.phpro.2010.01.102
24. Maraite, K.D.: Ein Beitrag zur Optimierung des Halbwarmziehens, Stahl und Eisen, Umformtechnische Band, vol. 13 (1988)

25. Kim, T.H., Kim, B.M., Choi, J.C.: Prediction of die wear in the wire-drawing process. J. Mater. Process. Technol. **65**(1–3), 11–17 (1997). https://doi.org/10.1016/S0924-0136(96)02235-2
26. Abe, K.: On the wire drawing lubricant. Trans. Japan Soc. Mech. Eng. **27**(181), 1439–1447 (1961)

Cycloidal Gearbox Model for Transient Analysis Implemented in Fortran with Constant Time Step 2nd Order Integrator

Roman Król(✉)

Kazimierz Pulaski University of Technology and Humanities, ul. Stasieckiego 54, 26-600 Radom, Poland
r.krol@uthrad.pl

Abstract. In this paper, the transient model of the cycloidal gearbox implemented in Fortran is presented. The constructed model of the gearbox is based on multi-body dynamics analysis and guarantees a 10-times faster kinematical and dynamical solution than the model designed in the engineering software (MSC Adams). The subroutines for contact modelling were implemented with the Kelvin-Voigt model. Opposite to the MSC Adams (constant contact stiffness), the contact in the Fortran model utilizes the curvatures of the contacting parts and their material parameters in each iteration of the analysis. Forces acting on external and internal sleeves have been compared with the MSC Adams model. Both models resulted in similar values of the forces acting on the sleeves. In the Fortran model, the 2nd order Runge Kutta method in the integration subroutine was used with the constant integration steps. The model in MSC Adams uses the solution method with an adaptive time step. Various dissipative models, which use initial contact velocity and the coefficient of restitution were tested without convergence. The convergence in the Fortran model was obtained only for the constant contact damping coefficient. The Fortran model allows for accurate results and can be further used in the optimization process.

Keywords: Cycloidal gearbox · Multibody dynamics analysis · Contact modelling · Output torque · Gearbox vibration

1 Introduction

Cycloidal gearboxes are widely applied in the drives of robotic systems, in the winches of rescue helicopters or offroad vehicles. The cycloidal gearbox design analyzed in this paper is characterized by the vibration of the output shaft with significant amplitude. The vibrations are caused by the variable angular velocity of the output shaft.

Contemporary research in cycloidal gearbox engineering concern the construction of the discrete models, which contain springs and dampers [1, 2] or modelling the cycloidal gearbox in the engineering software [2, 3]. In the iterative analysis, the current loading state of the cycloidal gearbox depends on the previous loading cycles. Excluding back iterations, the application of the discrete models for the modal and dynamic analysis

A. Puchalski et al. (Eds.): ICTD 2022, ACM 21, pp. 63–74, 2023.
https://doi.org/10.1007/978-3-031-31719-4_7

gives low accuracy results. To analyze the influence of the various contact models on the torque at the output shaft the transient models should be constructed based on multibody dynamics. Time courses of the forces and moments obtained in the analysis can be used in the development of fault detection methodologies and optimization. Current research in this area includes fault detection of the planetary drives [4–6], while the works, which concern cycloidal gearboxes among this studies are very rare.

In this paper, the transient model for fast analysis of the cycloidal gearbox, programmed in Fortran is described. It can further be used in the development of fault detection methods, vibration analysis or analysis of the contact models.

2 The Models of the Cycloidal Gearbox

Two models of the cycloidal gearbox were built. The first was designed in the MSC Adams software and the second was programmed based on Multibody Dynamics in Fortran. The second model was also implemented in Matlab software. Unfortunately, the adaptive time step allowed analysis to be very time-consuming. The version implemented in Fortran, which is a fast structural programming language with computer algebra facilities, uses the 2nd order Runge-Kutta method with constant integration steps. Both models consist of input and output shafts, two cycloidal wheels, 16 external sleeves and 8 internal sleeves (Fig. 1). All bodies are mounted on the bearings, which allow relative rotational motions between the parts. The parameters of bodies are given in Table 1. Loads are given in Table 2.

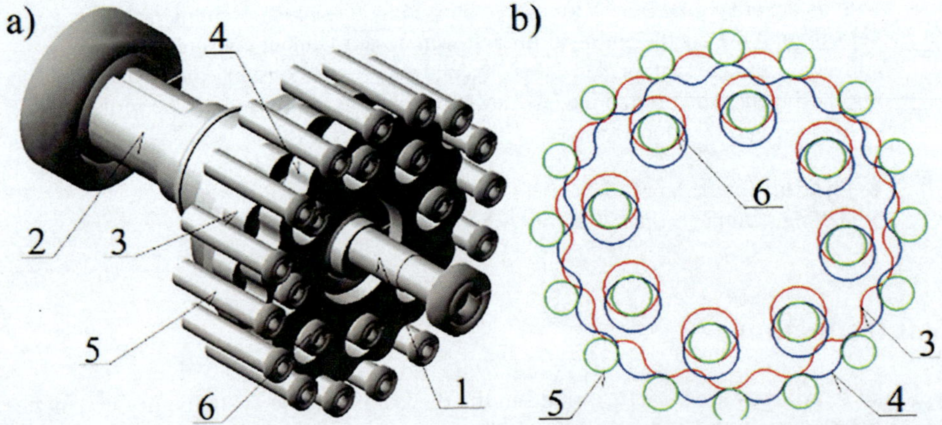

Fig. 1. Cycloidal gearbox models: designed in MSC Adams (a) and programmed in Fortran (b), where: 1 – input shaft, 2 – output shaft, 3 – internal cycloidal wheel, 4 - external cycloidal wheel, 5 – external sleeve, 6 – internal sleeve

Table 1. The masses and moments of inertia relative to the local coordinate systems. The local coordinate systems are placed in the parts' centre of gravity. The parts are rigid bodies

Body	Model parameters	
	Mass [kg]	Moment of inertia [kg·m^2]
Input shaft	0.2341	$1.846 \cdot 10^{-6}$
Output shaft	1.6345	$1.373 \cdot 10^{-4}$
Cycloidal wheel	0.5998	$1.57538 \cdot 10^{-4}$
Internal sleeve	0.048	$1.1013 \cdot 10^{-5}$
External sleeve	0.048	$1.1013 \cdot 10^{-5}$

Table 2. Loads and excitations in the models

Entity	Value
Angular velocity of the input shaft [Hz; rad/s; RPM]	8.33; 52.34; 500
Torque applied at the output shaft [Nm]	22

3 Multibody Dynamics Methods Used in the Model Programmed in Fortran

In each step of the integration subroutine, the following system of Eqs. (1) is solved. The model contains 28 bodies and 29 joints, one of which is a driver-joint and the others are revolute joints. The joints appear in the system of Eqs. (1) in the form of constraint equations, which are components of the Jacobian matrix. The method of Jacobian matrix formulation is described in [7, 8].

$$\begin{bmatrix} M & -D^T \\ D & 0 \end{bmatrix} \begin{bmatrix} \ddot{u} \\ \lambda \end{bmatrix} = \begin{bmatrix} F \\ \gamma \end{bmatrix}, \tag{1}$$

where: M – mass matrix (contains masses and moments of inertia of the parts), D - Jacobian matrix, 0 – zero matrix, $\ddot{u}$ – solved accelerations, λ – solved reaction forces, F - vector of applied forces and moments, γ – right-hand side vector of the acceleration equations.

In the described model the body-coordinate formulation was used. The constraint equations for the revolute joint and the driver-joint are given respectively in (2) and (3). Jacobian matrix satisfies (4), where the components of this matrix depend on the kind of joint. Submatrices of the Jacobian matrix for the revolute joint are given in (5), (6) and for the driver-joint in (7).

$$\Phi_{rev} = r_{P1} - r_{P2} = 0, \tag{2}$$

where: Φ_{rev} – revolute joint constraint equation, r_{P1}, r_{P2} – position vector of the revolute joint attachment point on the 1st and 2nd body respectively. The position vectors are given in the global coordinate system.

$$\Phi_{driver} = \varphi - f(t) = 0, \tag{3}$$

where: Φ_{driver} – driver-joint constraint equation, φ – the initial angle of a relative rotation of the connected parts, f(t) – time course of the rotation angle between the connected parts (i.e. f(t) = ωt for the simulation of the motion with the constant angular velocity ω).

$$D = \frac{\partial \Phi}{\partial u}, \tag{4}$$

where: D – Jacobian matrix, Φ – constraint equation, u – vector of body coordinates.

$$D_1 = \begin{bmatrix} -1 & 0 & -\breve{s}_{1X} \\ 0 & -1 & -\breve{s}_{1Y} \end{bmatrix}, \tag{5}$$

$$D_2 = \begin{bmatrix} 1 & 0 & \breve{s}_{2X} \\ 0 & 1 & \breve{s}_{2Y} \end{bmatrix}, \tag{6}$$

where: D1, D2 – Jacobian submatrix corresponding to the 1st and 2nd body connected to the revolute joint respectively, $\breve{s}$ – vector orthogonal to the joint attachment point local vector, relative to the centre of mass, X, Y – coordinates of the orthogonal vector.

$$D_{\text{drv}} = \begin{bmatrix} 0 & 0 & 1 \end{bmatrix}, \tag{7}$$

where: D_{drv} – Jacobian submatrix corresponding to the driven body. The right-hand side acceleration equations vector contains 0 in the components corresponding to the driver-joint. For the revolute joint, the right-hand side acceleration vector is solved according to (8).

$$\gamma = \dot{\breve{s}}_1\dot{\phi}_1 - \dot{\breve{s}}_2\dot{\phi}_2, \tag{8}$$

where: $\dot{\phi}_1$, $\dot{\phi}_2$- angular velocities of the 1st and 2nd body attached to the revolute joint.

4 Contact Modelling

Methods presented in the previous chapter were used to model three groups of parts in the cycloidal gearbox: 1) input shaft with the cycloidal wheels, 2) output shaft with the internal sleeves and 3) external sleeves. To transfer the moment from the input shaft to the output shaft, contact should be established between mentioned parts. The Kelvin-Voigt contact model based on the Hertzian theory with energy dissipation was programmed in the four subroutines, which check the contact between two cycloidal wheels, external sleeves and internal sleeves. Subroutines which check the contact between the cycloidal wheels and the external sleeves solve cycloidal wheel curvature in the contact point based on circular approximation from the three neighbouring points. The subroutines, which check the contact between the cycloidal wheels and the internal sleeves use a constant stiffness coefficient because contact surfaces have a constant curvature radius.

In the contact checking subroutines the cycloidal wheels and its holes were discretized. Convergence was obtained for the 600 (coarse analysis) and the 4000 (fine analysis) points in each cycloidal wheel and 20 (coarse analysis) and 100 (fine analysis) points in the cycloidal wheel holes in which the internal sleeves occur.

In the contact checking subroutine, the list of points is created, which distance to the centre of the sleeve is less than the sleeve's radius. From the list of points being in contact the point, which is closest to the sleeve centre is selected. Two points: W and Q detected by the contact checking subroutine are shown in Fig. 2.

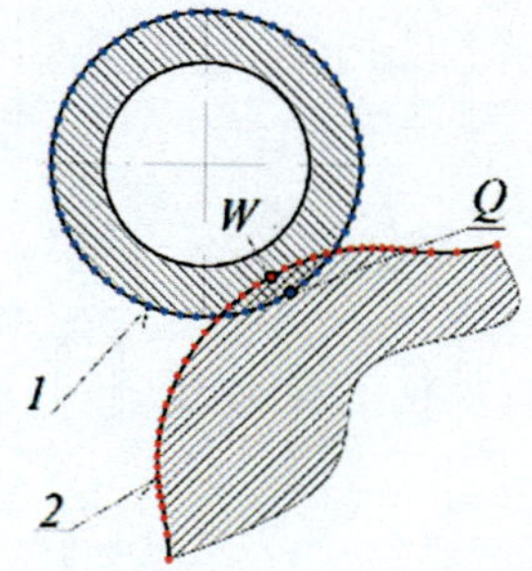

Fig. 2. Two points: W and Q detected by the contact checking subroutine. 1 – external sleeve, 2 – cycloidal wheel, Q – a point on the external sleeve, W – a point on the cycloidal wheel

Given the external sleeve centre of mass coordinates and point W (Fig. 2) coordinates, the contact normal vector is computed according to (9)

$$\vec{n} = \left[W_x - P_{nX}, W_y - P_{nY}\right], \tag{9}$$

where: W_X, W_Y – coordinates of the selected cycloidal wheel point being in contact with the external sleeve, P_{nX}, P_{nY} – coordinates of the given external sleeve centre of mass. To solve the friction force, the tangent contact vector (10) is solved using rotation matrix R_{90} (11).

$$\vec{t} = R_{90} \cdot \|\vec{n}\|, \tag{10}$$

$$R_{90} = \begin{bmatrix} \cos(90°) & -\sin(90°) \\ \sin(90°) & \cos(90°) \end{bmatrix}, \tag{11}$$

The vector r_{pQ} of contact point Q position relative to the global coordinate system can be computed from (12). The linear velocity $\dot{r}_{cW}$ of the point W can be computed from Eq. (13) with the cycloidal wheel rotation matrix specified in (14)

$$r_{pQ} = r_p + R_{exsl} \cdot \|\vec{n}\|, \tag{12}$$

$$\dot{r}_{cW} = \dot{r}_c + \dot{A}_c \cdot s_{cW}, \tag{13}$$

$$A_c = \begin{bmatrix} \cos(\varphi_c) & -\sin(\varphi_c) \\ \sin(\varphi_c) & \cos(\varphi_c) \end{bmatrix}, \tag{14}$$

where: r_p – vector of the external sleeves' centre of mass localization relative to the global coordinate system, R_{exsl} – radius of the external sleeve, $\|\vec{n}\|$ – contact normal vector with normalized length. A_C – rotation matrix of the cycloidal wheel, φ_c – angle of cycloidal wheel rotation, s_{cW} – local vector of the contact point W position relative to the cycloidal wheel centre of mass.

The normalized contact normal velocity is given in (15). Linear velocity (16) of the Q point can be computed by differentiating Eq. (12) and substituting (15). Contact normal velocity and contact tangential velocity can be solved using Eqs. (17) and (18) based on (9), (10), (13) and (16)

$$\left\|\dot{\vec{n}}\right\| = \dot{r}_{cW} - \dot{r}_p, \tag{15}$$

$$\dot{r}_{pQ} = \dot{r}_p + R_{exsl} \cdot \left\|\dot{\vec{n}}\right\|, \tag{16}$$

$$v_n = \left(\dot{r}_{pQ} - \dot{r}_{cW}\right) \cdot \left\|\dot{\vec{n}}\right\|, \tag{17}$$

$$v_t = \left(\dot{r}_{pQ} - \dot{r}_{cW}\right) \cdot \|\vec{t}\|, \tag{18}$$

where: v_n – contact normal velocity, v_t – contact tangential velocity. Theory, which concerns velocity computation is discussed in detail in [9, 10].

The contact normal velocity is used for the computation of the contact normal force (19) and the contact tangential velocity is used for the computation of the friction force (20). Contact normal force can be computed using various methods presented in [10]. In the Fortran model, the models of Lee and Wang or Herbert and McWhannel [10] were implemented. Unfortunately, the convergence for the specified models, which are designed for sphere-to-sphere contact and use initial contact velocity, was not obtained. Instead, the approach from MSC Adams documentation [11, 12] was used and the Heaviside function was utilized in the modelling of the contact stiffness and contact damping (19). The Heaviside function S(t – t0) (21) is linear in the range from 0 to t0.

$$F_n = S(\delta - 0.00001) \cdot K \cdot \delta^{1.5} + S(\delta - 0.00001) \cdot C \cdot v_n, \tag{19}$$

$$F_t = -\mu \cdot S(v_t - 0.00005) \cdot F_n, \tag{20}$$

where: S – Heaviside step function (21), K – contact stiffness, C – contact damping, v_n - contact normal velocity, v_t – contact tangential velocity, μ – static coefficient of friction, δ – penetration depth (the distance between the W and Q points).

$$S(t - t_0) = \begin{cases} 0 & for\ t < 0 \\ \frac{1}{t_0} \cdot t\ for\ t \in (0, t_0) & . \\ 1 & for\ t > t_0 \end{cases} \tag{21}$$

The stiffness coefficient (22) is solved based on the surface curvature of the contacting bodies and the parameters given in (23). The contact stiffness coefficient was multiplied by the H multiplier. The values of the H multiplier and contact damping C are presented in Table 3.

$$K = \frac{4H}{3\pi\left(h_c + h_p\right)}\sqrt{\frac{R_c \cdot R_p}{R_c \pm R_p}}, \tag{22}$$

$$h_c = \frac{1 - \nu_c^2}{\pi \cdot E_c}, h_p = \frac{1 - \nu_p^2}{\pi \cdot E_p}, \tag{23}$$

where: K – contact stiffness, H – contact stiffness multiplier, R_c – radius of curvature of the cycloidal wheel surface, R_p – radius of curvature of the external sleeve surface, ν_c, ν_p – Poisson ratios of the cycloidal wheel and the external sleeve material, E_c, E_p – Young modulus of the cycloidal wheel and the external sleeve respectively. The sign in the denominator between the R_c and R_p values depends on the contact with the cycloidal wheel pit or the cycloidal wheel lobe (the negative or positive radius of curvature of the cycloidal wheel).

Table 3. Parameters used in contact modelling

Parameters	Values used by contact detection subroutines (external and internal sleeves)
ν_c, ν_p	0.3
E_c, E_p [N/m^2]	$2{\cdot}10^{11}$
C [Ns/m]	8
H	0.01

Computed values of the forces and moments acting on the external sleeves, internal sleeves and cycloidal wheels are substituted to the system of Eqs. (1) in the F vector in each iteration of the 2nd order Runge-Kutta algorithm. In the solution process, the constant integration step was used with 10^{-5} s. The analysis time was set to 0.6 s.

5 Results of the Analysis in the Fortran Model

One of the benefits of the transient model programmed in Fortran is the possibility to analyze a graph of every entity in the simulation process. MSC Adams allows the user to enter a constant value of the contact stiffness, while in the Fortran model the value of the contact stiffness depends on the curvature of contacting bodies at the point of contact. Contact stiffness at the moment of contact between the cycloidal wheel and one of the external sleeves is shown in Fig. 3. Time courses of the input torque, output torque and forces acting on the internal and external sleeves are shown in Fig. 4. The output torque applied at the output shaft as a load is constant, but due to the geometry of the cycloidal gearbox in the analysis, the time courses of the input and output torques are oscillating.

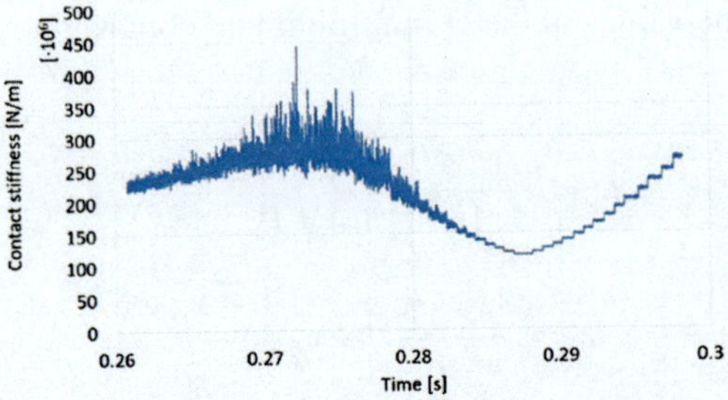

Fig. 3. Contact stiffness in the time range, when contact between the external sleeve and cycloidal wheel occur

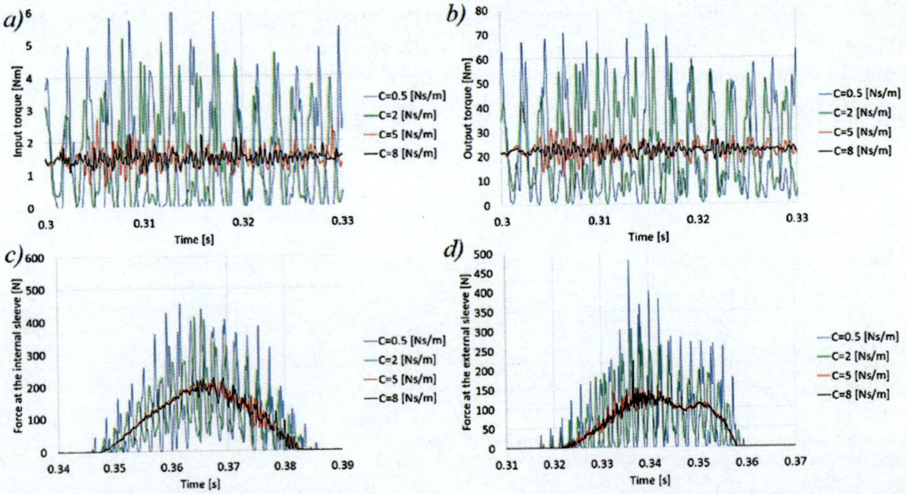

Fig. 4. Physical entities measured in the model programmed in Fortran: torque at the input shaft (a), torque at the output shaft (b), the force acting on the internal sleeve (magnitude) (c), the force acting on the external sleeve (magnitude) (d)

A comparison of the force acting on the internal sleeve and the external sleeve obtained with coarse discretization (600 points for each cycloidal wheel) and fine discretization (4000 points for each cycloidal wheel) are shown in Fig. 5 (a, b). The data obtained in MSC Adams were compared with the data from the transient model programmed in Fortran in Fig. 5 (c, d).

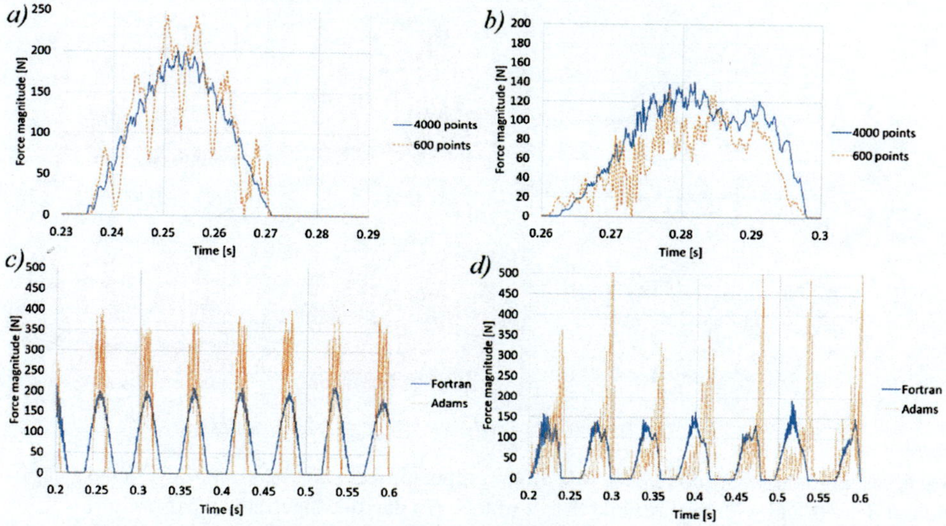

Fig. 5. Results computed in the model implemented in Fortran for a various number of points in the cycloidal wheel (600 points and 4000 points): force acting on the internal sleeve (magnitude) (a), the force acting on the external sleeve (magnitude) (b). Results from MSC Adams and the model implemented in Fortran: force acting on the internal sleeve (magnitude) (c), the force acting on the external sleeve (magnitude) (d)

6 Application of the Fortran Model in Fault Diagnosis

The presented Fortran model can be used in fault diagnosis of the cycloidal gearboxes. Analyses for the gearbox without defects and the gearbox with artificial lobe defects (Fig. 6a) were performed. The parametric shape of lobe was cut by the linear function connecting two points on the cycloidal wheel, which correspond to the beginning and ending angular values. The angle α (Fig. 6a) appears as an argument in the parametric equations of the cycloidal wheel. The analyses were performed with the 0.0003 m backlash to allow gearbox movement without mechanism locking.

The magnitude square coherence function (24) (Fig. 6 b-e) was solved for the output torques computed in the gearbox with and without artificial defects of the lobe.

$$C_{xy}(f) = \frac{|G_{xy}(f)|^2}{G_{xx}(f)G_{yy}(f)}, \tag{24}$$

where: $C_{xy}(f)$ – magnitude square coherence as a function of the frequency, $G_{xx}(f)$, $G_{yy}(f)$ – auto-spectral densities of the output torque, $G_{xy}(f)$ – cross-spectral density of the output torques, x(t) – time course of the output torque in the gearbox without defects, y(t) – time course of the output torque in the gearbox with the artificial lobe defect, f – frequency.

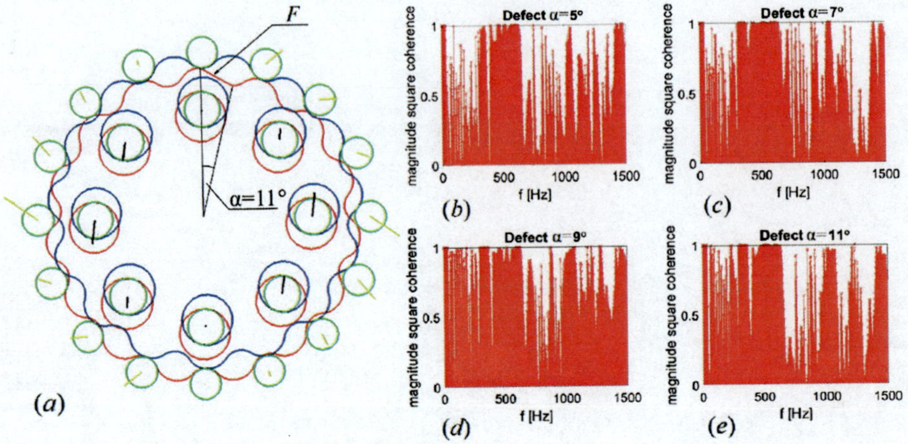

Fig. 6. Magnitude square coherence (b-e) of the output torques calculated in the cycloidal gearbox without defects and with the artificial defect of lobe (F) due to exagerrated wear (a)

7 Discussion of Results

The presented model implemented in Fortran offers a short analysis time, which allows using this model in the optimization processes. For coarse discretization, the solution time is 4 min, while for fine discretization it is 25 min. Analysis results from MSC Adams presented in Fig. 5 (c, d) require more than one hour for the solution process.

In the model implemented in Fortran, various contact force models [10] were tested, but convergence was not obtained. These models were designed for two contacting spheres, while the contact in the analyzed cycloidal gearbox is cylinder to cylinder. The contact stiffness solved according to the Hertzian model should be multiplied by the coefficient $H = 0.01$ to satisfy convergence conditions.

Analysis for various damping coefficients (Fig. 4 a-d) shows that contact damping has a significant influence on the quality of results. For low contact damping values, high oscillations emerge, while higher values of contact damping guarantee smooth results. The MSC Adams results presented in Fig. 5 (c, d) show significant oscillations in the force diagrams for the internal and external sleeves. It can arise from the improper value of the contact damping coefficient, which was set to the same value in the model implemented in Fortran. Analysis of the 3D model in MSC Adams requires another set of contact parameters (stiffness, damping and the penetration depth at which contact damping coefficient reaches maximum value).

To use the model implemented in Fortran in the optimization process, it is beneficial to reduce the solution time of the single simulation. Unfortunately, it can be seen from Fig. 5 (a, b), that application of the discretization, which is too coarse leads to inaccurate results. The results for the 600 points of discretization show significant oscillations of the force acting on the internal sleeve in Fig. 5 (a). The level of discretization depends on the optimization aims.

The presented Fortran model can be used in the fault diagnosis of the cycloidal gearbox. The coherence function should be equal to 1 in the healthy gearbox, and analyses for the artificial defect of the lobe for $\alpha < 5°$ did not detect any disturbances. For higher angular values of the defect, Fig. 6e shows a more coherence decrease in the range of 600–1000 Hz than Fig. 6b for the lower artificial lobe defect. Obtaining an exact interpretation of the coherence functions requires further studies and analyses. The output torque in the cycloidal gearbox is a random function which oscillates around the mean value relative to the load and gearbox ratio. The influence of the previous iterations and contact modeling allows output torque to have characteristics of the random time course.

8 Conclusions

Summing up the results, the model implemented in Fortran guarantees a fast solution and can be successfully used in the optimization process. The discretization of the model should be taken into account when the accuracy of the optimization results is considered. The contact stiffness should be multiplied by the experimentally adjusted coefficient to allow convergence of the solution process. Contact damping coefficient can significantly influence the accuracy of the results.

This research utilised the MSC Software National Scientific Software License, operated by the TASK Computer Centre in Gdańsk (Poland). This license was funded by a computational grant obtained by Kazimierz Pulaski University of Technology and Humanities in Radom, Poland.

References

1. Wikło, M., Król, R., Olejarczyk, K., Kołodziejczyk, K.: Output torque ripple for a cycloidal gear train. Proc. Inst. Mech. Eng. C J. Mech. Eng. Sci. **233**(21–22), 7270–7281 (2019). https://doi.org/10.1177/0954406219841656
2. Król, R.: Resonance phenomenon in the single stage cycloidal gearbox. Analysis of vibrations at the output shaft as a function of the external sleeves stiffness. Arch. Mech. Eng. **68**(3), 303–320 (2021). https://doi.org/10.24425/ame.2021.137050
3. Król, R.: Kinematics and dynamics of the two stage cycloidal gearbox. AUTOBUSY – Technika, Eksploatacja, Systemy Transportowe **19**(6), 523–527 (2018). https://doi.org/10.24136/atest.2018.125
4. Plöger, D.F., Zech, P., Rinderknecht, S.: Vibration signature analysis of commodity planetary gearboxes. Mech. Syst. Signal Process. **119**, 255–265 (2019). https://doi.org/10.1016/j.ymssp.2018.09.014
5. Lei, Y., Han, D., Lin, J., He, Z.: Planetary gearbox fault diagnosis using an adaptive stochastic resonance method. Mech. Syst. Signal Process. **38**(1), 113–124 (2013). https://doi.org/10.1016/j.ymssp.2012.06.021
6. Wang, T., Han, Q., Chu, F., Feng, Z.: Vibration based condition monitoring and fault diagnosis of wind turbine planetary gearbox: a review. Mech. Syst. Signal Process. **126**, 662–685 (2019). https://doi.org/10.1016/j.ymssp.2019.02.051
7. Nikravesh, P.E.: Planar Multibody. Dynamics. (2018). https://doi.org/10.1201/b22302
8. Nikravesh, P.E.: Planar multibody dynamics: Formulation, programming and applications (2007)
9. Xu, L.X., Chen, B.K., Li, C.Y.: Dynamic modelling and contact analysis of bearing-cycloid-pinwheel transmission mechanisms used in joint rotate vector reducers. Mech. Mach. Theory **137**, 432–458 (2019). https://doi.org/10.1016/j.mechmachtheory.2019.03.035

10. Flores, P., Lankarani, H.M.: Contact Force Models for Multibody Dynamics, vol. 226. Springer, Cham (2016). https://doi.org/10.1007/978-3-319-30897-5
11. MSC Software: "MSC Adams Solver Documentation"
12. MSC Software, "MSC Adams View Documentation."

Experimental Setup for Nondestructive Testing of Composite Structures Using Laser Spot Thermography

Michał Sobczak(✉), Adam Machynia, Ziemowit Dworakowski, and Jakub Roemer

Department of Robotics and Mechatronics, AGH University of Science and Technology, Krakow, Poland
msobczak@agh.edu.pl

Abstract. Laser thermography is a sub-domain method in active thermography that uses a laser as the heat source. The method allows for precise control of the beam shape, pulse duration, and delivered energy. Therefore, this technique can be applied especially for delicate structures where testing conditions must be maintained precisely. The following work includes the introduction of a laser spot thermography technology demonstrator that was designed and constructed at AGH UST. The paper presents laser thermography laboratory experiments along with the developed procedure for automatic defect identification. The measurements were used to establish an automatic procedure and algorithm for automatic defect identification based on regression methods and a K-means clustering algorithm. The procedure leads to the indication of faulty and flawless regions of the sample. Experiments were conducted on a carbon fiber reinforced polymer (CFRP) sample with artificially created defects in the form of Teflon inserts imitating delaminations on various depths.

Keywords: Laser spot thermography · delamination detection · non-destructive testing · automatic defect identification

1 Introduction

In the face of constantly increasing requirements for the safety and reliability of the structures, the development of safety measures becomes necessary to ensure the structure's full operational performance. Nowadays, the usage of lightweight materials for high-performance applications does not allow for high safety factors as it was done in the past. To ensure structural safety of structures non-destructive testing (NDT) methods should be applied and developed.

1.1 Non-destructive Testing Methods

Non-destructive testing methods do not alter the state of the object and give insight into its health and condition. They could be performed on-demand or

A. Puchalski et al. (Eds.): ICTD 2022, ACM 21, pp. 75–84, 2023.
https://doi.org/10.1007/978-3-031-31719-4_8

in scheduled intervals of time and provide precise information about the state of the inspected object. They are a group of local methods, that are performed usually for part of the object. All of the methods widely used nowadays, have already appropriate inspection procedures standardized and are successfully used and applied in NDT. However, the development of new materials requires the development of new methods for specific applications and specific materials.

1.2 Thermography

The non-destructive testing method that offers vast inspection possibilities is thermography. It is a study of obtaining and analyzing thermal information from non-contact thermal imaging devices. Its main advantage is that it is contactless, so the inspections can be conducted in harsh environments without danger to the operator. Moreover, the data is two-dimensional and is obtained in real-time. Two main branches of thermography can be distinguished. The first branch is passive thermography which does not use any source of excitation. In this case, the thermal image of the object is taken and thermal patterns are analyzed. It is a widely known method for inspecting buildings [2,4], and searching for defects in photo-voltaic systems [5].

The other thermography branch, called active thermography, utilizes heat excitation and analyzes patterns of heat flow. Active thermography finds application in non-destructive testing. The excitation can be in form of optical excitation, mechanical, or other such as e.g. Eddy currents [1]. The method that uses mechanical excitation is called vibrothermography [10]. In that case, the heat is generated by friction between the unbounded material layers. The vibrations are generated at the sample by a piezo-electric transducer (PZT). The signal generated for PZT can be in form of a lock-in or chirp signal.

The inspection procedure can be performed either in form of a point-by-point scan [6] (or one single area scan [12]) or in the form of a continuous scan [9], where the camera and heat source are moved with respect to the surface of the inspected sample.

Optical excitation can have two different sources. The classical approach uses lamps as an excitation. Halogen lamps [7] are most widely used for that purpose, but also flash lamps [13] can be applied. Lamp excitations are preferable for inspections of large surfaces. The undeniable advantage of halogen lamps is the low price of the heat source. On the other hand, the operator usually has to deal with non-uniform heat distribution and sometimes with reflections caused by the lamps. The usage of flash lamps eliminates problems with longer response times. In the case of active thermography, a few methods of inspections are possible, among others: Lock-in [14], Pulse [3], and Pulse-Phase [8] (Fig. 1).

1.3 Laser Spot Thermography

Another heat source utilizes a laser to provide energy to the sample. The laser excitation gives the possibility to precisely control inspection parameters such as delivered energy and length of the laser pulse. Moreover, the laser spot generates

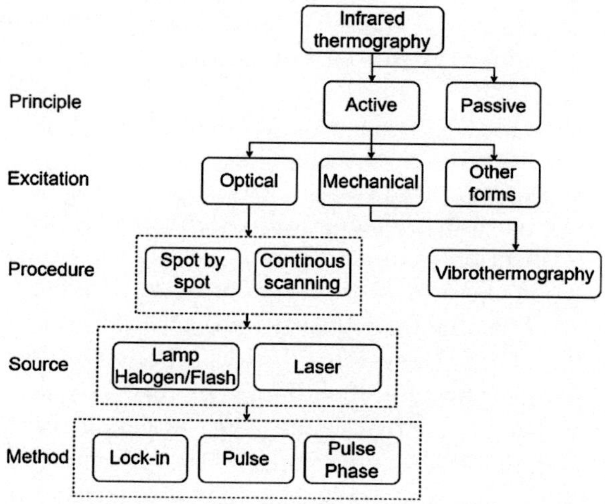

Fig. 1. Infrared thermography classification.

a spherical heat wave that gives the possibility to detect cracks that are perpendicular to the tested surface [11]. Lamps, on the other hand, generate a planar heat wave that penetrates tested material and allows only for the detection of delaminations and cracks parallel to the surface. Laser thermography is especially suitable for small samples and precise inspections. It generates measurements with better spatial resolution, which gives a possibility to detect smaller defects. The drawback of the method is a longer inspection time, due to the smaller area of a single inspection point and a greater number of points on the measurement grid. One of the challenges of laser thermography is to compose the data in order to create a consistent image of the inspected sample.

2 Motivation for Research

Our work concentrates on laser spot thermography which works on the same basis as pulse thermography in the classical approach with lamp excitation. This method is not new but has recently been gaining attention due to development in inspection equipment, significantly lower prices of thermographic cameras, and their greater possibilities. The additional challenge for laser spot thermography is the lack of strict inspection procedures, which will be developed within the scope of our project. Inspections with halogen lamps using methods such as pulse thermography are easy to implement, as only the lamp must be synchronized with the camera. Laser thermography, however, requires precise synchronization of laser, camera, and manipulators responsible for the position inspected sample. Additionally, the data from consecutive inspection points must be gathered and merged to create an image of the whole scanned area. In this article test rig for laser spot thermography that is capable of providing repeatable measurements is presented along with framework for data processing leading to automatic defect detection.

3 Test Rig

For the purposes of laser spot thermography, the experimental setup has been developed. Setting inspection parameters and starting the procedure is done on a PC computer. It is the part that works as a human-machine interface. PC computer is equipped with National Instruments LabVIEW software. PC is connected to the main controller (1) of the system which is a PXIe-8821 computer with NI Linux Real-Time operating system. Real-Time Operating System (RTOS) allows for precise synchronization of control signals, such as signals responsible for starting the laser beam, beginning the acquisition, and the movement of the positioners. Thanks to the precise signals' synchronization the taken measurements are repeatable and have a similar form which influences the quality of processed results. Part of the system responsible for creating a laser beam consists of the CS412 Amtron laser controller (2) and Limo laser source. The Limo laser source model 120-F400-DL808-AV5-A of power 120W generates a wave with 808nm length. The applied controller gives the possibility to set the power and shape of the impulse signal. The system's operating temperature is ensured by the Termotek P308 cooling system. The laser beam is propagated in fiber optics to the LPH-LD80-R-35-0 Limo laser head (6) with collimating lenses that generate a laser spot of size approx. 10 mm on the sample. The temperature field measurement is taken with FLIR A655SC infrared camera (3) that operates in the 7.5–15 μm wavelength range. The camera works with a resolution of 480 × 640 pixels and is able to record sequences up to 50 frames per second. The precise positioning of the inspected sample (5) is ensured by the Standa 8MTL1401-300 - Direct Drive Linear Translation Stages (4) controlled by the SPiiPlusCMNT motor controller. Two positioners are connected orthogonally allowing for movement in the X-Y plane. The working area of the stand is 300 × 300 mm (Figs. 2 and 3).

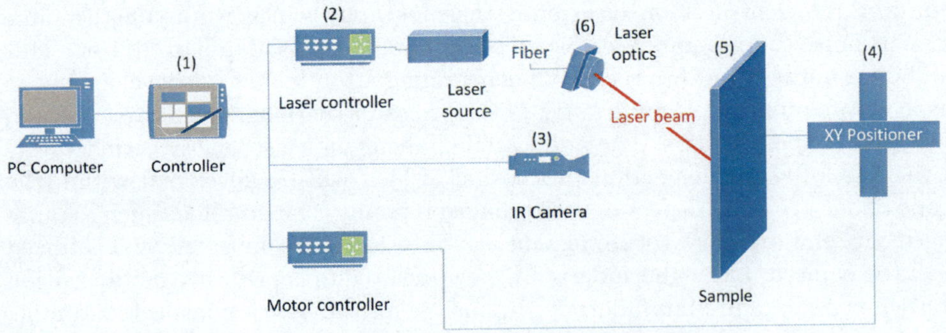

Fig. 2. Scheme of the experimental setup.

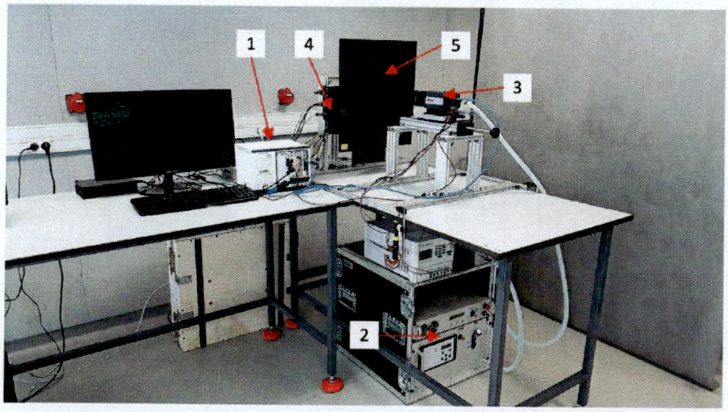

(a) Experimental setup.

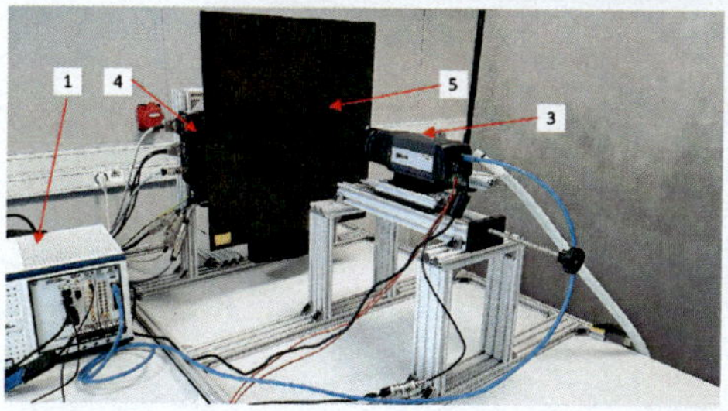

(b) Experimental setup - camera on the platform.

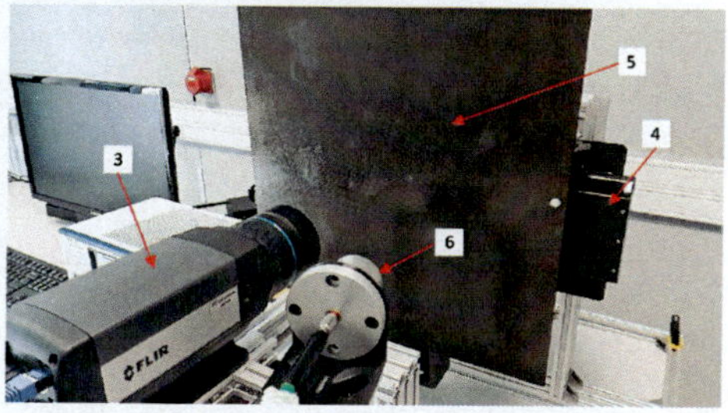

(c) Experimental setup - camera and laser head.

Fig. 3. Photographs of the test stand. (1) PXIe Controller, (2) laser controller, (3) FLIR infrared camera, (4) Standa linear positioners, (5) composite sample, (6) laser head.

4 Experimental Procedure

4.1 Tested Sample

The experiment has been performed on Carbon Fiber Reinforced Polymer (CFRP) plate prepared from 985-GF-3070PW prepreg material. The sample consisted of ten layers of material aligned in the $[0°/90°/0°/90°/0°]$s pattern. The initial dimensions of the sample were $500 \times 450 \times 2$ mm. Due to the limited workspace of the experimental stand, the dimensions of the scanned area were set to 200×160 mm. Artificial defects were introduced to the sample in form of Teflon inserts. Defects imitate delaminations in the composite and they were placed on different depths of the sample. The inspected area covers four defects with diameters of 20 mm and 30 mm, placed between the first and second layers and between the third and fourth layers (Fig. 4).

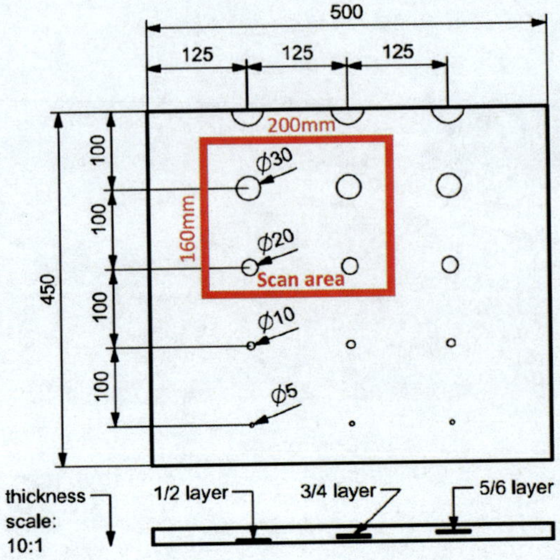

Fig. 4. Experimental sample dimensions.

4.2 Scanning Path

The inspection points were distributed every 5 mm resulting in a grid consisting of 33×41 measurement points. Overall 1353 measurements were taken. In order to avoid the thermal influence of neighboring points inspection path was not aligned as a straight line. Measurements were taken every 2 points in the X direction and every 3 points in the Y direction. A changed path allowed for heat dissipation from the neighborhood of the previous point on the path. For each measurement point, 97 J of energy was delivered and the pulse length was set to 800 ms. The camera shutter speed was set to 20 fps. For each point, 244 frames were recorded, which resulted in 12.2 s of obtained thermographic sequence (Fig. 5).

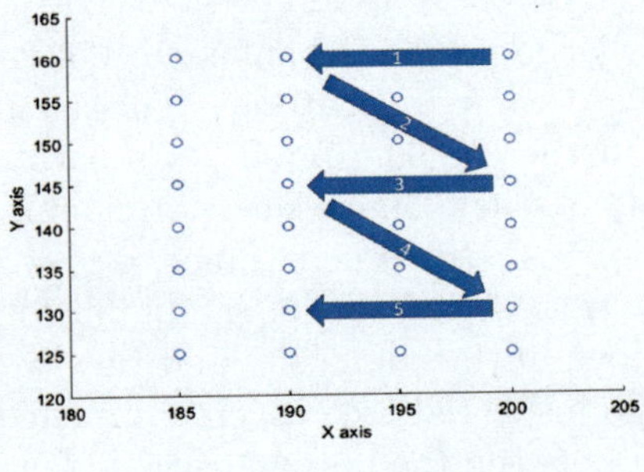

(a) First part of the path.

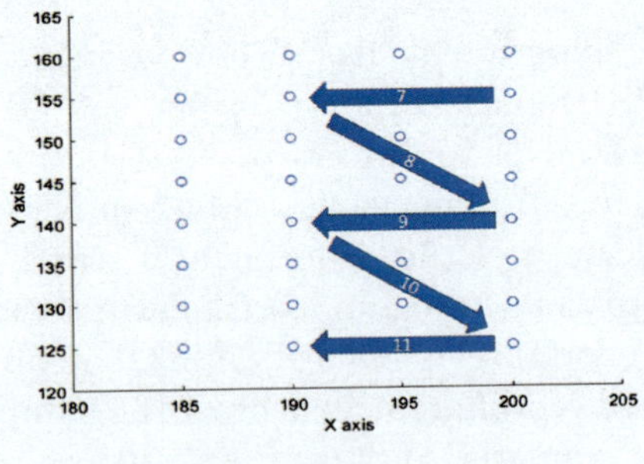

(b) Next part of the path.

Fig. 5. Inspection path.

5 Defect Detection Procedure

5.1 Defect Detection Algorithm

For the purposes of automatic defect detection based on gathered thermographic data, a data processing algorithm was developed. Firstly, the position of the region of interest (ROI) is determined. For this purpose reference frame is chosen – the frame near the peak of temperature, that is, the transition from the heating into the cooling process. The selected frame is binarized with an automatic threshold utilizing the Otsu method, so the resulting image indicates the region directly heated with the laser. The position of the center of the spot is obtained as a centroid of the binary object. Found point coordinates are set as a center of a rectangle ROI of size $w \times h$ for all images in the sequence. In this research, the ROI size is experimentally set to $w = h = 80$ pixels. The next step leads to determining a background as an averaged set of frames before the excitation. The mean temperature is calculated for the frames from 2nd up to 5th at the beginning of the sequence. The very first frame is rejected as it consists of noise. Next, in order to focus on the impact of laser excitement, the background is subtracted from all the following frames. Further, for consecutive frames, from the 20th frame up to 100th frame, the mean values of the ROI temperature are calculated. This stage of processing is shown in Fig. 6.

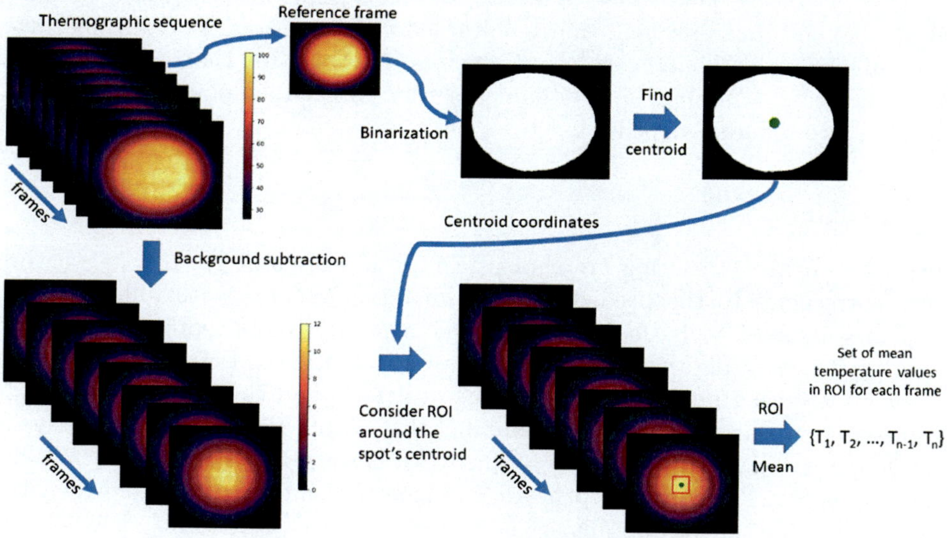

Fig. 6. Scheme of data processing algorithm.

Having time history data of mean temperature for each measurement point curve fitting is performed. The least squares algorithm is used. At this stage, the amount of data is reduced from the whole thermographic sequence to four parameters precisely describing the curve. Data fitting is performed only on the part of the curve that describes the cooling stage of inspection, which starts at the 20th frame. The equation that describes the curve is:

$$y = ae^{-bx} + ce^{-dx}$$

with starting point (1,1,1,1), and assumption that $a < c$. Curve fitting is demonstrated in Fig. 7.

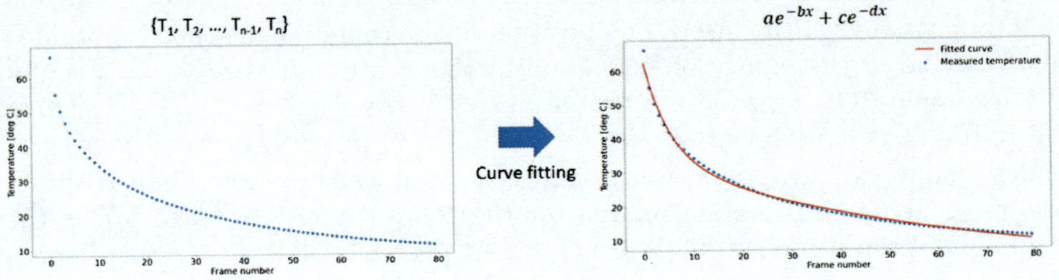

Fig. 7. An example of proposed curve fitting.

To assess the state of the inspected sample unsupervised machine learning technique was used. Due to curve fitting, the data point is represented by four parameters and this representation is suitable for data clustering. K-means clustering algorithm with division into 3 clusters was used for that purpose. The choice of clusters number is reasonable as there are intact points and defects placed at two different depths.

5.2 Results

The results of the clustering are shown in Fig. 8a. Points in green color on the figure correspond to the predicted group of intact points. With different colors points assumed to be faulty are marked, separate colors represent different defects in terms of depth. For results assessment the ground truth is prepared. As some measuring points are at the edge of defects and cover both intact and damaged area, they are marked as unlabeled. Compared to the ground truth, no false positive indications were obtained. All defects were correctly localized. Defects placed on the same depth are also clustered into the same clusters. In Fig. 8b confusion matrix for this set is presented. False positive rate is equal to zero and also false negative rate is very low.

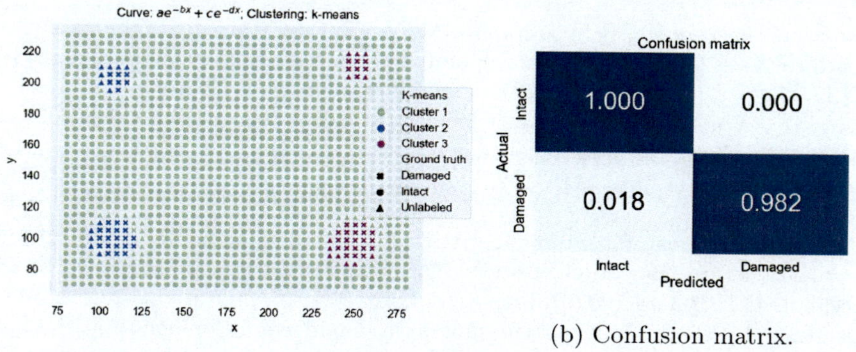

(a) K-means clustering results.

(b) Confusion matrix.

Fig. 8. Results.

6 Conclusions

In the study, the system for defect detection using laser spot thermography has been developed along with the data processing procedure for automated defect detection. The experimental setup consisted of a PC computer, a PXIe computer, a laser controller, an infrared camera, and two linear positioners. The usage of a real-time operating system allows for precise synchronization of the generated control signals. The experiment performed on the CFRP sample provided satisfactory results with no false-positive values and only one false negative. Unsupervised learning methods were used to assess the state of the sample at the measuring point. The data processing algorithm starting with ROI detection and curve fitting results in a reduction in thermographic data from the whole thermographic sequence into four parameters describing the curve. In the next step, k-means clustering is used to divide data into 3 clusters. Results are presented on the graph. In the future, more experiments will be performed on composite materials, and more data processing modules will be developed.

Acknowledgment. The research has been financed within the scope of project no. 0001/L-11/2019 "Laser thermography testing for damage detection in composite structures" financed by the National Center for Research and Development, Poland. Part of the work was also financed by the AGH University of Science and Technology, WIMiR, research grant no. 16.16.130.942/GD/2022 and a subsidy from the Department of Robotics and Mechatronics.

References

1. Barakat, N., Mortadha, J., Khan, A., Abu-Nabah, B., Hamdan, M., Al-Said, S.: A one-dimensional approach towards edge crack detection and mapping using eddy current thermography. Sens. Actu. A: Phys. **309**, 111999 (2020). https://doi.org/10.1016/j.sna.2020.111999

2. Barreira, E., Almeida, R.M., Moreira, M.: An infrared thermography passive approach to assess the effect of leakage points in buildings. Energy Build. **140**, 224–235 (2017). https://doi.org/10.1016/j.enbuild.2017.02.009
3. Feng, Y., Yin, C., Huang, X., Chen, K., Dadras, S.: Rapid defect detection for spacecraft in infrared reconstructed images based on automatic optimum mosaic algorithm, pp. 1718–1723 (2021). https://doi.org/10.23919/ACC50511.2021.9483232
4. Garrido, I., Lagüela, S., Otero, R., Arias, P.: Thermographic methodologies used in infrastructure inspection: a review-data acquisition procedures. Infrared Phys. Technol. **111**, 103481 (2020). https://doi.org/10.1016/j.infrared.2020.103481
5. Kandeal, A., et al.: Infrared thermography-based condition monitoring of solar photovoltaic systems: a mini review of recent advances. Solar Energy **223**, 33–43 (2021). https://doi.org/10.1016/j.solener.2021.05.032
6. Li, T., Almond, D., Rees, D.A.S., Weekes, B.: Crack imaging by pulsed laser spot thermography. In: Journal of Physics: Conference Series, vol. 214, p. 012072 (2010). https://doi.org/10.1088/1742-6596/214/1/012072
7. Luo, Q., Woo, W.L., Yang, Y.: Temporal and spatial deep learning network for infrared thermal defect detection. NDT & E Int. **108**, 102164 (2019). https://doi.org/10.1016/j.ndteint.2019.102164
8. Maldague, X., Largouët, Y., Couturier, J.P.: A study of defect depth using neural networks in pulsed phase thermography: modelling, noise, experiments. Revue Générale de Thermique **37**(8), 704–717 (1998). https://doi.org/10.1016/S0035-3159(98)80048-2
9. Montinaro, N., Cerniglia, D., Pitarresi, G.: Flying laser spot thermography for the inspection of aerospace grade fibre metal laminates, pp. 206–210 (2017). https://doi.org/10.1109/MetroAeroSpace.2017.7999565
10. Pieczonka, L., Aymerich, F., Brozek, G., Szwedo, M., Staszewski, W., Uhl, T.: Modelling and numerical simulations of vibrothermography for impact damage detection in composites structures. Struct. Control Health Monit. **20**, 626–638 (2013). https://doi.org/10.1002/stc.1483
11. Rodríguez-Aseguinolaza, J., Colom, M., González, J., Mendioroz, A., Salazar, A.: Quantifying the width and angle of inclined cracks using laser-spot lock-in thermography. NDT & E Int. **122**, 102494 (2021). https://doi.org/10.1016/j.ndteint.2021.102494
12. Saeed, N., King, N., Said, Z., Omar, M.: Automatic defects detection in cfrp thermograms, using convolutional neural networks and transfer learning. Infrared Phys. Technol. **102**, 103048 (2019). https://doi.org/10.1016/j.infrared.2019.103048
13. Zhou, J., Du, W., Yang, L., Deng, K., Addepalli, S., Zhao, Y.: Pattern recognition of barely visible impact damage in carbon composites using pulsed thermography. IEEE Trans. Ind. Inf. **18**(10), 7252–7261 (2022). https://doi.org/10.1109/TII.2021.3134184
14. Ólafsson, G., Tighe, R., Boyd, S., Dulieu-Barton, J.: Lock-in thermography using miniature infra-red cameras and integrated actuators for defect identification in composite materials. Optics Laser Technol. **147**, 107629 (2022). https://doi.org/10.1016/j.optlastec.2021.107629

Neural Network-Based Road Surface Condition Monitor Concept

Krzysztof Dobaj(✉)

Cracow University of Technology, Warszawska 24, 31-155 Cracow, Poland
kdobaj@pk.edu.pl

Abstract. The paper presents the monitoring of the road surface condition identification. The monitor is based on artificial neural networks. They were used both to filter the test results and to determine the pavement condition appropriate for the previously filtered measurement data. The input quantities to the formulated monitor were: pressure in the brake caliper and longitudinal wheel slip. The results of the tests constituting the braking processes carried out with different dynamics and different vehicle loads were used. A concept for determining the condition of the road surface relied on the neural network estimation of the Pacejka Magic Formula road parameters. The neural network for this application was created using data obtained from the structural simulation model of the vehicle braking process. This model used the description of the wheel-road cooperation given by the Magic Formula Model and the design parameters of the car braking system. The obtained accuracy of the pavement condition monitor was considered sufficient from the point of view of further works. They will be consisting in the use of the formulated monitor in the algorithms of continuous diagnostics of braking systems.

Keywords: Road Surface · Estimation · Neural Networks · Vehicle Braking System · Wheel Slip

1 Introduction

In currently produced motor vehicles, control of the braking system comes down to periodical control of the electrical efficiency in the mechatronic parts of these systems. The real- time (in driving conditions) control of the effectiveness of the mechanical part of the braking systems is not carried out. The methodology for determining road surface condition, presented in the paper, has a specific application. It is designed to work in cooperation with the diagnostic monitor for the current diagnostics of the whole braking system. The electrohydraulic control of braking systems requires the controllers, where the algorithm of the real- time brakes diagnostics can be implemented. Increasing the accuracy of the electrohydraulic braking systems controlling (for example in hybrid, electric and autonomous cars) requires increasing the diagnostics accuracy of these systems. When the diagnostics is based on the real- time braking system monitors, there is a necessity to increase the accuracy of determining the road surface conditions by these monitors. The method of identifying the road surface condition formulated in the paper uses artificial neural networks. They were built using the Matlab software.

A. Puchalski et al. (Eds.): ICTD 2022, ACM 21, pp. 85–94, 2023.
https://doi.org/10.1007/978-3-031-31719-4_9

2 Analysis of the Currently Developed Concepts for Estimating Road Adhesion in Real Time of Vehicle Movement

The issues of continuous estimation of road adhesion in vehicle motion conditions are undertaken by many authors. In the work [4], the analysis of vibrations in the frequency domain was conducted. Attention was drawn to difficulties in determining the road adhesion coefficient in conditions of low acceleration affecting the drive system and the steering system of the vehicle. A vehicle with an electric motor in the wheel hub was analyzed. The road adhesion coefficient was determined by analyzing the longitudinal and transverse dynamics of the vehicle. Its dependence on the frequency of vibrations in the wheel guidance system was determined, using the measurement of the operating current of the engine driving the vehicle and the speed of rotation of the steering column. Time-frequency analysis methods were used.

Another way to estimate the current value of the road adhesion coefficient is to use the Extended Kalman Filter. The work [6] discusses its use, combined with the use of the APF filter. These methods are used to determine the sideslip angle of the front wheels. Then, the values of the adhesion coefficient were determined based on the formulated iterative algorithm. This algorithm is based on the analysis of the stabilizing moment of force, which strongly depends on the tire sideslip angle. This ensures a quick and accurate evaluation of the adhesion coefficient in real time. The method has been tested experimentally and by simulations using Carsim and Matlab Simulink.

In the work [7], a method of estimating the value of the longitudinal tire- road adhesion coefficient was proposed along with the corresponding value of the circumferential slip of the wheel, which is the maximum value that can occur on a given road surface. Determining this value is crucial for improving the control of anti-lock braking systems of the vehicle. The proposed approach, in order to ensure adequate accuracy, assumes the classification of the obtained measurement data into groups related to different tire- road adhesion. For this purpose, machine learning methods were used. In order to minimize classification errors, boundary conditions were formulated. In turn, the algorithms seeking the maximum of the longitudinal adhesion coefficient were formulated as recursive algorithms based on the Kalman filter. This algorithm is based on measuring changes in the rotational speed of the wheel and changes in its dynamic radius while the vehicle is moving. The proposed concept was verified by simulation in Matlab.

The classification approach supporting the determination of the most accurate value of the adhesion coefficient, using fuzzy logic, is presented in [8]. The coefficient of tire-road adhesion was estimated separately for the conditions of acceleration, braking and cornering. Then, using data fusion methods, the share of these traffic states in complex maneuvers performed while driving was adjusted. A vehicle model with three degrees of freedom was formulated. They concerned the vehicle displacement in the longitudinal and transverse directions as well as its rotation in relation to the vertical axis. The model formulated by Pacejka was used to describe the behavior of the tire. Using the described model assumptions, the proposed method was verified by simulations. Then, the simulation results were experimentally verified by performing road tests. These tests included the measurement of the longitudinal and lateral acceleration of the car as well as the steering wheel angle and were carried out on surfaces with different adhesion conditions. The verification confirmed the sufficient accuracy of the method. However, it was indicated that the rapidly changing states of vehicle motion – sudden, alternating braking and acceleration, combined with fast turns, can significantly reduce the accuracy of the formulated method.

In the work [9], a method for determining the current road adhesion coefficient was introduced on the basis of a 3-axis accelerometer located in the inner part of the tire. The measurement of accelerations allowed for the analysis of the longitudinal and transverse dynamics of the tire. The coefficient of longitudinal and transverse adhesion was determined on the basis of the determined forces at the contact of the wheel with the road and the tire sideslip angle. These values were determined on the basis of accelerations measured with a 3-axis accelerometer placed on the inner part of the tire. Test drives were carried out using various surfaces. The results of the analysis were consistent with the results of standardized measurements.

In the works analyzed by the author on the issues of diagnostics of the braking system in real time, there is no emphasis on the precise identification of the surface condition [2, 3]. For example, in the work [2], the approximated dependence of the car deceleration on the pressure in the brake system is used to assess the effectiveness of the braking system. The road adhesion conditions are taken into account only by categorizing them into sufficient and insufficient for the real-time braking system diagnostics. The adhesion conditions are considered as insufficient when the wheel locks (or the ABS is activated) before the minimum deceleration value required for the braking system to be considered operational. This approach is correct from the point of view of simplifying the diagnostic process, however, it does not ensure the high accuracy of the braking system diagnostics, which would be ensured by a more detailed diagnosis of road adhesion. Increased accuracy of the braking system assessment, and thus the reaction to slight loss of braking efficiency, seems to be of key importance due to the increased effectiveness of driver support systems based on the braking system. In turn, in [3] the operation of the introduced brake diagnostics system for typical adhesion coefficients for the most common road conditions was checked (by simulating it).

The work, carried out by the author, focuses not only on the estimation of road adhesion conditions, but also on the impact of the accuracy of their estimation on the process of real-time brake diagnostics. The work is a preliminary analysis in which the focus was on the development of a methodology for recognizing whether road adhesion will or will not allow for the diagnosis of brakes in real- time driving conditions. Dry or wet asphalt was considered as the condition of the road allowing for the diagnosis of the brakes. On the other hand, the condition of the roadway for which brake diagnostics is impossible is assumed to be a snowy or icy roadway. The advantage of such a rough approach is that there is no need for additional car equipment. We use sensors that are part of the ABS / ESP systems that are mandatory in modern cars - brake pressure sensor and sensors. This reduces the costs of implementing the discussed algorithm in motor vehicles.

3 Road Tests and Their Results Approximation

The analysis of determining road adhesion was made on the basis of various braking runs on a dry road, which were to reflect the operational braking runs - i.e. different vehicle loads, different initial braking speeds and different braking dynamics (pressure changes in the braking system over time). The used braking processes are:

- Process 1: initial vehicle speed: 80 kph, tire pumping pressure 2,0 bar, mass of the research vehicle with driver, high braking dynamics
- Process 2: initial vehicle speed: 80 kph, tire pumping pressure 2,0 bar, mass of the research vehicle with driver and 300 kg load, medium braking dynamics
- Process 3: initial vehicle speed: 50 kph, tire pumping pressure 2,0 bar, mass of the research vehicle with driver and 300 kg load, low braking dynamics
- Process 4: initial vehicle speed: 30 kph, tire pumping pressure 2,0 bar, mass of the research vehicle with driver and 300 kg load, medium braking dynamics.

During the above-mentioned braking processes, the linear speed of the car was measured (with the Correvit optical head), the speed of the car wheels (impulse-rotary transducer) and the pressure in the brake system (strain gauge). The measurement apparatus was installed in the Skoda Octavia passenger car used in the research.

The formulated surface condition monitor was based on the analysis of the dependence of the pressure in the brake caliper and the wheel slip. The wheel slip is defined as:

$$s = \frac{v - \omega_{wheel} \cdot r_{dwheel}}{v} \quad (1)$$

where:

- r_{dwheel} vehicle wheel dynamic radius,
- ω_{wheel} wheel angular speed,
- ω_{wheel} vehicle linear speed.

The vehicle speed, used in the above equation can be calculated by the Antilock Braking System algorithm [10].

One wheel was considered - left front. Implementation of the formulated model must include the analysis of the behavior of all the wheels of the vehicle with the braking pressure supplied to each of them. Consequently, the monitor must take into account the vehicle's braking force distribution or, in mechatronic braking systems, it must read the braking pressure values from sensors at each wheel.

The first step of the road surface determination algorithm was the approximation of the measurement results. For this purpose, a two-layer neural network was used with a sigmoidal activation function of neurons in the hidden layer and a linear neural activation function in the output layer (Fig. 1). There was conducted a comparison of the data approximation efficiency by neural network, working in different braking runs. For this purpose, a selection of 100 points of the dependence wheel slip-brake fluid pressure was made for a given braking process. And sorted in ascending order with respect to the slip. The data prepared in this way, randomly divided into the training, validation and test set in proportions 70/15/15, respectively, was used to formulate the

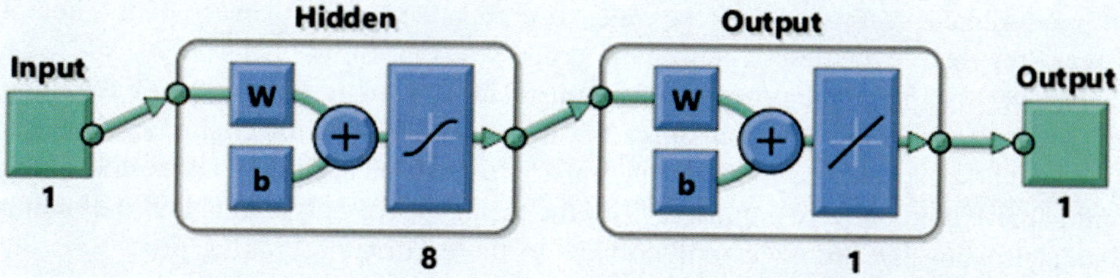

Fig. 1. Diagram of a neural network used for approximation of measurements [9]

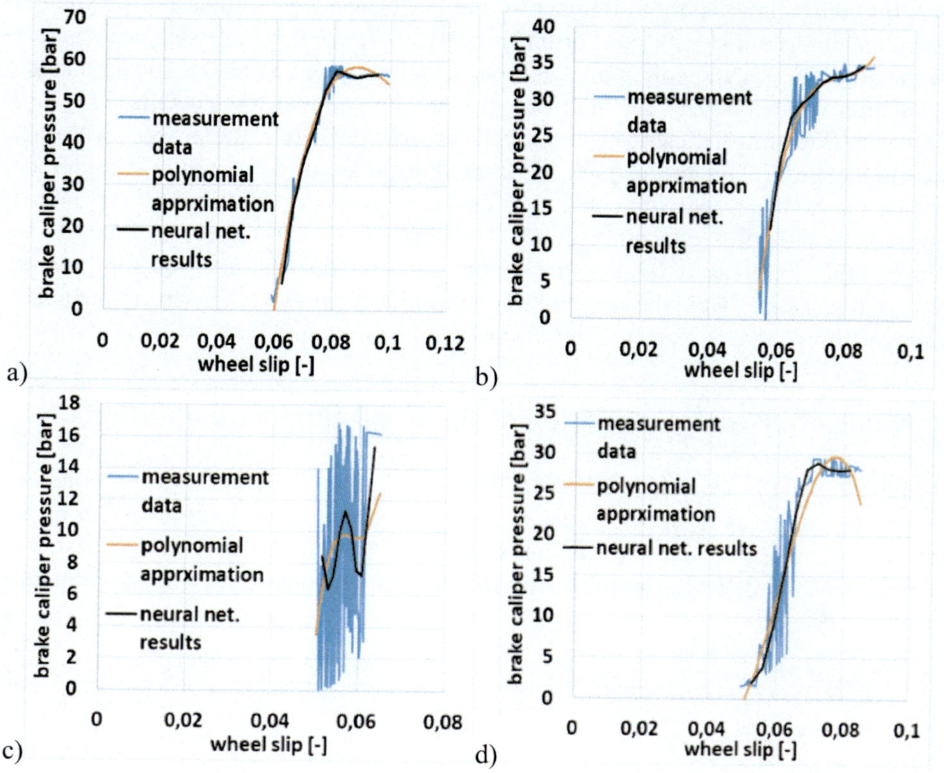

Fig. 2. The measurement data neural approximation: a) braking process 1, b) braking process 2, c) braking process 3, d) braking process 4

neural network. The input values were the wheel slip, on the basis of which the pressure values required to obtain them were determined by the neural network (output values). The formulated neural network was then used to estimate the pressure values for wheel slip values that were not present in the data used to formulate the neural network. These values were chosen in the manner which divided the area between the smallest and the largest slip values in a given braking run into equal parts. The discussed formulation of the neural network was repeated for each of the 4 previously discussed braking runs. The discussed unification of the data size used to formulate the network and to approximate the measurement results made it possible to compare the effectiveness of the neural network for each of the described braking processes (Fig. 2).

The wheel slip unit [-] means that it is dimensionless. In the case of the first braking process, we have relatively small spread in the brake caliper- wheel slip dependence. These conditions provide a good next neural approximation accuracy. The results of the neural approximation were compared with the results of the polynomial approximation conducted using the 3rd order polynomial. In the end of the braking process 1, the performance of the polynomial approximation is worse than the performance of the neural approximation. The polynomial approximation was not used in the further work, involved with road surface estimation, due to its tendency of losing the details. In the second braking process, the difference in the performance of neural and polynomial approximation is not significant. For this process at area of large pressure scatter (for a slip of 0.06 to 0.07), the polynomial approximation provides a smaller spread than the neural approximation. This fact confirms that the polynomial loses details. This truth is also confirmed in the case of the braking process 3. This is the process with high braking dynamics. Due to this fact, the limited accuracy of the measuring equipment provide the higher wheel slip- braking pressure dependency scatter. The polynomial approximation did not reflect this scatter. The braking process 4 also has an increasing nature of the slip-pressure relationship. In this case, the dispersion of the slip-pressure relationship caused that in the range of slip values from 0 to 0.07 the slope of the neural approximation curve is greater than the slope of the polynomial approximation curve.

4 Neural Network Estimation of the Road Surface Conditions

The next step of the analysis is the neural determination of the road adhesion conditions for sets of 10 points of slip-pressure relationship, obtained as a result of the neural approximation discussed in the previous chapter. In order to determine the adhesion of the road, a neural network formulated on the basis of data obtained using the following formula was used:

$$p(s) = \frac{r_{dwheel} \cdot F_z \cdot D \cdot \sin[C \cdot arctan\{B \cdot s - E(B \cdot s - \arctan(B \cdot s))\}]}{\mu_{pad-disc} \cdot A_{piston} \cdot r_{disc}} \tag{2}$$

where:

- r_{dwheel} vehicle wheel dynamic radius,
- F_z wheel vertical load,
- *B, C, D, E* Magic Formula model coefficients,
- $\mu_{pad\text{-}disc}$ brake pad- brake disc friction coefficient,
- A_{piston} surface of the brake caliper piston,
- r_{disc} radius of the brake disc.

The Eq. 2 results from the insertion of an expression defining the force at the contact of the wheel with the roadway formulated by Pacejka (Magic Formula) into the equation of motion of the braked wheel. To obtain data formulating the neural network, data involved with 4 variants of the road surface condition was used. This was done using the coefficients in the Table 1 below.

Table 1. The Magic Formula coefficients for typical road surface conditions [1]

	B	C	D	E
dry asphalt	10	1,9	1	0,97
wet asphalt	12	2,3	0,82	1
snow	5	2	0,3	1
ice	4	2	0,1	1

It has been computationally shown that the wheel inertia resistance in the processes of passenger car braking has values not exceeding 10 Nm, constituting about 2% of the braking torque value. On the other hand, the moment of rolling resistance takes values two orders of magnitude smaller than the moment of the resultant longitudinal force at the contact of the wheel with the road. Due to the illustrative nature of this work, the discussed moments of resistance were omitted when formulating the structural model. The vehicle weight distribution, the geometric parameters of the braking system and the brake pad-brake disc friction coefficient used for the analysis concerned the front wheel of the Skoda Octavia 1.6 car. The nominal value of the block-brake disc friction coefficient (equal to 0.27) was adopted on the basis of the analysis performed in [5].

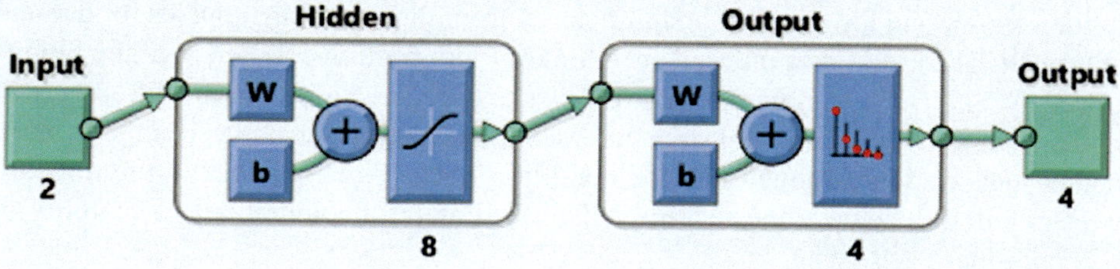

Fig. 3. Diagram of the neural network used to estimate road grip [9]

The used neural network (Fig. 3) was matching a given data sample (2 elements - slip and pressure) to 4 values of the Magic Formula model coefficients. It was a two-layer network with a sigmodal function of activating neurons in the hidden layer and a normalized exponential function of activating neurons in the output layer (softmax function). For each of the 4 performed braking runs, an analysis was performed involving the use of various data variants used to formulate the neural network. These variants included determining the pressure in the braking system (using formula No. 1) for the slip values from a given braking run and for the slip value with an imposed step of its change of 0.0015, the same for each braking run. For each of these slip ranges, two braking pressure datasets were determined. The first was involved with the pressure, required to obtain the slip values, when the friction coefficient of the braking linings is about 0.27. It was the determined, nominal value of the friction coefficient. The second braking pressure dataset included pressure values, required for achieving the demanded slip values when the braking linings friction coefficient is reduced by 50% from the nominal value. The data used to formulate the neural network for a specific braking run included 400 data samples - 100 for each of the four pavement conditions. The list of data variants is as follows:

- variant 1: slip values from the tests, the brake is operational;
- variant 2: slip values obtained from tests, brake weakened by 50%;
- variant 3: slip values with imposed step, the brake is operational;
- variant 4: slip values with imposed step, brake weakened by 50%.

For each of the four above variants, the neural network determined the coefficients of the Pacejka model for 10 samples of slip-pressure data resulting from the neural approximation of the data described in the previous chapter for each of the four braking runs. Therefore, we have a total of 16 variants of the neural network analyzed. The application of the imposed step of changing the slip value speeds up the calculations. This acceleration is provided because there is no need for creating data for the formulation of the neural network each time after collecting the measurement data. It is enough to create the network once and use it for different braking runs, However in this approach, the nature of the input and output data differs to a greater extent than in the case of using the slip from the measurements. These conditions may deteriorate the accuracy of the neural network that estimates the road adhesion.

According to the Pacejka model, in the area of slip that does not activate the ABS, we have slight differences between the courses of the tire-road force a function of wheel slip for dry and wet asphalt. However, the relationships between the tire-road force and wheel slip for a snowy and icy road differ significantly from the according courses for dry and wet asphalt. The use of data on weakened brakes (with friction coefficient of the linings reduced by 50% from the nominal value) allows to check the impact of the differences in the tire- road force and the wheel slip courses on brake diagnostics. Specifying, to check whether they are large enough that the condition of the brakes, affecting the pressure values required to achieve the set slip values, will affect the neural network ability of road conditions classifying.

Table 2. The Mean values of the Magic Formula coefficients estimated via neural network in best and worst case

Best accuracy: Braking process 4, efficient brake, imposed slip step 0,0015			
B: 8,67	C: 2,12	D: 0,86	E: 0,99
Worst accuracy: Braking process 4, efficient brake, slip values from measurement			
B: 8,24	C: 2,25	D: 0,58	E: 0,99

In order to determine the effectiveness of the road condition (adhesion) assessment in the analyzed cases, the average values of the individual coefficients of the Pacejka model were compared with the values, which corresponded to dry asphalt. Based on the analysis, it can be concluded that the accuracy of our neural method of estimating road adhesion is 10% in the most favorable cases and 20% in the least favorable cases (Table 2). Theoretically, the least favorable case should be the case when the neural network is created using data with the imposed slip step and damaged brakes. This situation results from the fact that the road tests were performed with the brakes in nominal technical state (efficient condition) and the test slip values from the road tests differ from the imposed slip step. In practice, the worst mapping accuracy of the Magic Formula model coefficients occurs for the 4th braking process, for the input data for the neural network with the slip values from the tests and the nominal brake pad- disc friction coefficient. The best accuracy is also for the 4th braking process, for nominal brake state and for the slip values with an imposed step. This situation confirms the lack of regularity between the theoretical difficulty for the neural network and its real effectiveness (Fig. 4).

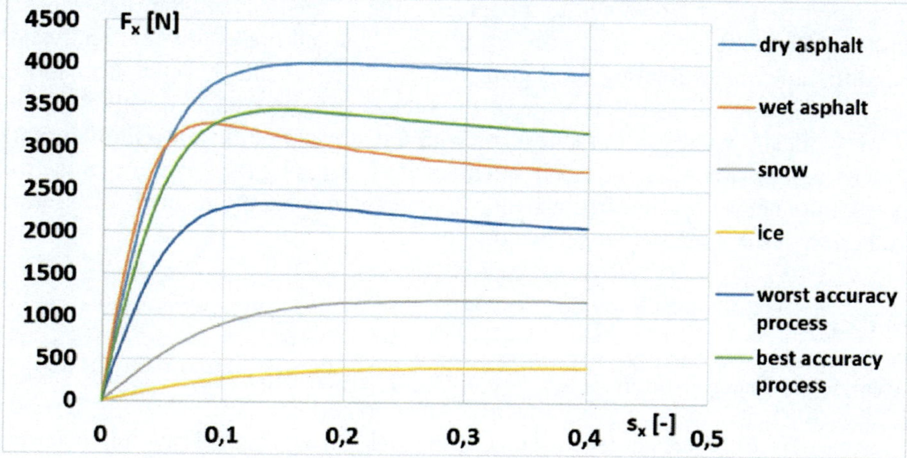

Fig. 4. Comparison of relationships between tire-road force (F_X) a wheel slip (s_X) for exemplary Magic Formula coefficients and for the averaged coefficients of the Magic Formula model obtained through the neural network

The figure above shows tire-road force courses in the function of wheel slip for typical values of the Magic Formula coefficients for 4 types of road surface conditions. The figure also depicts the relationships between the tire- road force and wheel slip for the mean values of the coefficients of the Magic Formula model for the braking process, in which the neural network estimated the Magic Formula coefficients most accurately and least accurately. The difference in the values of the tire- road force for the same slip in the case of the best and worst accuracy of the MF model coefficient mapping is about 1800 N. It is a significant value. However the tire- road force- wheel slip course of the worst Magic Formula coefficient mapping is above the tire- road force- wheel slip course for a snow-covered road. The difference between these courses is large. Due to this fact, on the basis of the formulated neural algorithm for estimating road adhesion, it is possible to roughly estimate the condition of the road adhesion- qualify it to the group that allows or disables brake diagnostics. For this purpose it is necessary to formulate a criterion of achieving the limiting slope of linear part of the tire- road force- wheel slip course. Limit of the slope determines the road adhesion as possible for braking system diagnostics. In order to determine the limit slope a series of simulation and experimental tests must be conducted. The test must explore the analyzed method of determining the adhesion together with the real- time diagnostic monitor of the brake system, which will be formulated further.

5 Conclusions

The analysis presented in the paper confirms that the formulated method of the road condition classification is promising for the future. It has been proved that for the linear ranges tire- road force- wheel slip course, the differences in the condition of the brakes by 50% in relation to their nominal condition affect the pressures required to obtain the specified values of wheel slip to a degree so much smaller than the road adhesion conditions. Due to this fact, it was proved that deteriorated brakes do not significantly cause difficulties in estimating road grip. The accuracy of the formulated method is about 10–20%, regardless of the state of the brakes and the dataset creating a neural network (with slip values from the tests or with a fixed step of change). Further works will include: supplementing the discussed method of assessing the condition of the road with a monitor of the real-time brake diagnostics and increasing the accuracy of assessing the adhesion of the road and brakes condition.

References

1. Garrido, P., Soares, F., Moreira, A.P. (eds.): CONTROLO 2016. LNEE, vol. 402. Springer, Cham (2017). https://doi.org/10.1007/978-3-319-43671-5
2. Tretsiak, D.: Experimental investigation of the brake system's efficiency for commercial vehicles equipped with disc brakes. Proc. Inst. Mech. Eng. Part D J. Auto. Eng. **226**(6), 725–739 (2011)
3. Chen, L., Luo, Y., Bian, M., Qin, Z., Luo, J., Li, K.: Estimation of tire-road friction coefficient based on frequency domain data fusion. Mech. Syst. Signal Process. **2017**(85), 177–192 (2017)
4. Gajek, A.: Możliwości diagnozowania układów przeciwblokujących pojazdów samochodowych Zeszyty Instytutu Pojazdów/ Politechnika Warszawska **2003**(2/49), 33–54 (2003)
5. Liu, Y.-H., Li, Y.-Y., Yang, X.-W., Wu, J.J.: Estimation of tire-road friction coefficient based on combined APF-IEKF and iteration algorithm. Mech. Syst. Signal Process. **2017**(88), 25–35 (2017)
6. Regolin, E., Ferrara, A.: SVM classification and Kalman filter based estimation of the tire-road friction curve. IFAC-PapersOnLine **2017**(1), 33382–3387 (2017)
7. Li, L., Yang, K., Jia, G., Ran, X., Song, J., Han, Z.-Q.: Comprehensive tire-road friction coefficient estimation based on signal fusion method under complex maneuvering operations. Mech. Syst. Signal Process. **2015**(56–57), 259–276 (2015)
8. Matsuzaki, R., Kazuto, K., Seki, R.: Intelligent tires for identifying coefficient of friction of tire/road contact surfaces using three-axis accelerometer. Smart Mater. Struct. **24**(2), 025010 (2015)
9. Materials from the Matlab software
10. Polkowski S.: Układy bezpieczeństwa i komfortu jazdy-informator Bosch, praca zbiorowa, WKŁ (2013)

Method of Fault Detection of Drive Systems Using Features of Park's Vector Hodograph

Krzysztof Roczek(✉) and Marek Fidali

Silesian University of Technology, Konarskiego 18A, 44-100 Gliwice, Poland
krzysztof.roczek@gmail.com, marek.fidali@polsl.pl

Abstract. Electric drives are popular in different branches of industry. This trend applies also to the press shops that are using stamping press lines equipped with induction or servomotors. Despite of using of high quality components, unexpected breakdowns take place what causes machine stop. In consequences very often, high costs are generated for the company. From that reason, it is recommended to use diagnostic systems that will allow predicting or early detecting the breakdown.

One of the possibility is application of the system based on Motor Current Signature Analysis (MCSA). The method is described in the standard ISO 20958. The document defines the characteristic frequencies for particular breakdowns that may be recognized on the electric current spectrum. Frequency analysis of stator current signal allow to define the peaks that will testify to the damage.

Deeper analysis with use of Park's vectors and Park's hodographs is also possible. In the article the authors presents the method of detection and identification of electrical drive system faults using features of Parks's hodographs extracted from them using methods of image analysis. Results of the research shows that relevant features are useful for diagnostics of electrical drive system faults.

Keywords: MCSA · Park's transform · EPVA · Stamping press · motor diagnosis

1 Introduction

Metal forming is commonly used in the automotive industry. The stamping process is carried out on the press lines, which could be treated as a critical industrial object due to extremely high costs of production losses caused by press line breakdown. The tandem press line [1] is a complex system where any fault could influence the whole stamping process and in some cases cause unexpected breakdown of the line. MFMEA analysis presented in [2] indicates that main drive system and especially motor is critical for reliability of the whole press line. The reasonable way of avoiding unexpected drive system breakdowns is application of Condition Based Maintenance (CBM) and Predictive Maintenance (PdM). Contemporary market offers solutions for press line condition monitoring. There are many possibilities of condition monitoring systems that can be applied also for motor of press drive system. Most of them perform either vibration or temperature analysis [3, 5, 6]. In addition, solutions based on MCSA (Motor Current

A. Puchalski et al. (Eds.): ICTD 2022, ACM 21, pp. 95–104, 2023.
https://doi.org/10.1007/978-3-031-31719-4_10

Signal Analysis) could be a valuable source of diagnostic information. Some commercial system offered on this area [4, 7, 8] are mainly based on evaluation of current signal in time and frequency domain. In the article described advantages of MCSA in motors condition assessment and presented original method of condition evaluation based on Parks vector trajectory images. During the research, we focused mainly on analysis of real life current signals of induction motor powered by inverter in contrast to most of research performed in laboratory conditions where motors were powered directly from the mains.

2 Electric Motors Condition Assessment Based on Current Analysis –State of the Art

There are many methods of electric current analysis to perform diagnosis of induction motor. One may list here:

– Spectral Analysis (FFT) [9, 10, 18]
– Motor Current Signature Analysis [11–21]
– Huang-Hilbert Transform and Empirical Mode Decomposition [26, 27]
– Extended Park's Vector Approach [16, 19, 20, 22, 23]
– Park-Clark Transform [18, 23, 24, 25]
– Time-frequency analysis- wavelet analysis [10]

2.1 Spectral Analysis

Frequency analysis is the most commonly used method of motor current signal processing [10]. It is because many of motor faults creates specific frequency components whose may be detected and identified in signal spectrum. Moreover, a standard that describes those frequencies was elaborated: ISO20958 [18]. This document indicates the methods of characteristic frequencies calculation for typical motors problems like eccentricity, problem rotor bar, rolling element bearing defect or short circuit [18]. Of course, the formulas are based on the number of poles and basic frequency of the power supply. It must be mentioned that spectral analysis is more accurate for steady- state applications. For the systems with variable load or with transient conditions, this method cannot be applied [10]. Analysis with use of frequency domain provide information about the existence of symptom at some frequency. In order to define if a fault exists it is also necessary to analyse the amplitude of a peak. One should also remember that if the frequency component can be observed in both spectra: voltage and current, then we have an electric problem. If the characteristic frequency component is visible only electric current the face with mechanical problem [14].

2.2 Motor Current Signature Analysis

This method is widely described by ISO standard 20958. The document [18] precisely describes the motor improper operation and explains the particular consequences in stator current. The complete method is based on analysis of stator current signatures. Many of Motor Current Signature Analysis (MCSA) experiments were performed for laboratory stations without frequency inverter. Unfortunately, this kind of applications can be rare meet in industrial application, what meets that results of those experiments cannot be fully utilized in production plants. Authors of [11] also utilizes the frequency domain analysis in order to identify the eccentricity problem.

The main method of MCSA used for the given analysis is based on Park's transform and hodographs. That kind of transformation changes the representation of 3-phase system into 2- dimensional one. Equation 1 presents the Park's transform utilized for the generation of hodographs.

$$\begin{bmatrix} d \\ q \end{bmatrix} = \frac{2}{3}\begin{bmatrix} cos\,\delta & cos\left(\delta - \frac{2\pi}{3}\right) & cos\left(\delta + \frac{2\pi}{3}\right) \\ sin\,\delta & sin\left(\delta - \frac{2\pi}{3}\right) & cos\left(\delta + \frac{2\pi}{3}\right) \end{bmatrix}\begin{bmatrix} u \\ v \\ w \end{bmatrix} \quad (1)$$

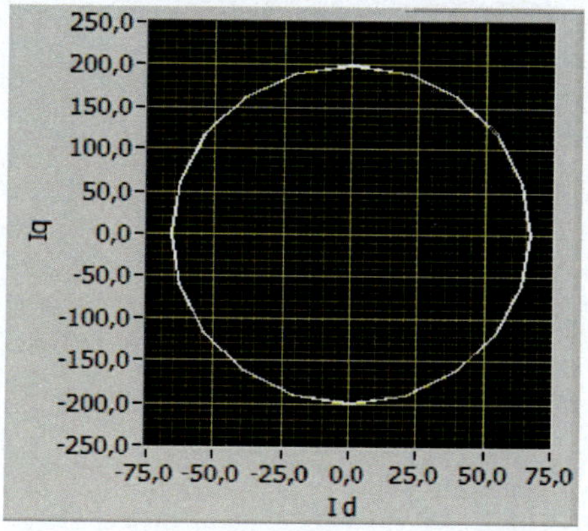

Fig. 1. Park's hodograph for three-phase system supplied from standard 50 Hz network power supply

Three-phase system can be then shown as rotational coordinate system (see Fig. 2) what creates hodographs (see Fig. 1 and Fig. 3). Ideal hodographs for the motor supplied directly with use of contactors (3× 400 VAC 50 Hz) will be a circle (see Fig. 1). Trajectory presented on Fig. 3 is related to the motor driven by frequency inverter. Each sharp tip is related to operation of one transistor in frequency inverter. Activation of a transistor causes charging of capacitor what is indicated as increase of the peak. Deactivation discharges the capacitor and lowers the electric current intensity value.

Therefore, Park's hodographs shape is strictly related to the operation of drive system and can be used for the purpose of machinery diagnosis.

Detection of faults based on Parks's hodographs could be performed using two ways. The first one takes into consideration analysis of the hodographs shapes and feature extraction in order to perform pattern recognition and classification of actual condition of diagnosed motor. The second method omits feature extraction step and consider hodograph as an pattern which could be recognized and classified exemplary by use of artificial neural networks. This approach requires enough amount of training and testing examples. Due to small data set, we focused on elaboration of hodographs

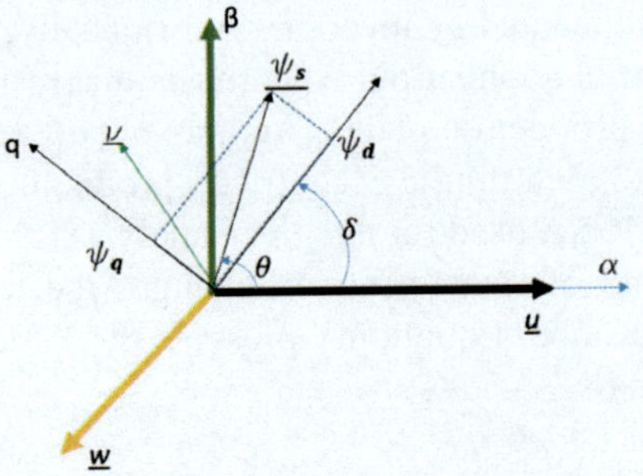

Fig. 2. Rotational coordinate system

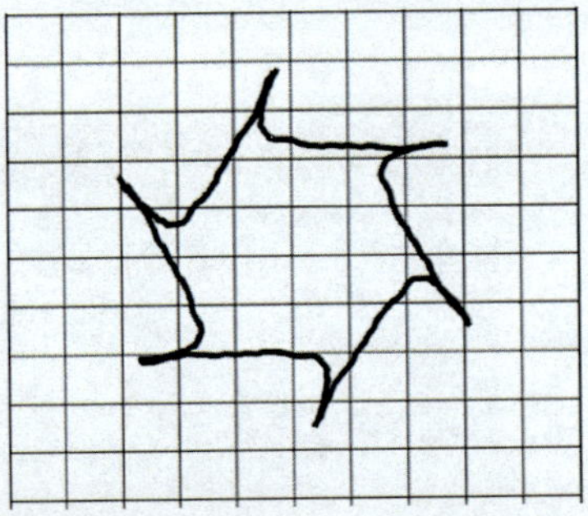

Fig. 3. Example of Park's hodograph

analysis method to extract features suitable for resolving classification problem using simple classifier based on Euclidian measures and leave-one-out validation.

3 Measurement System and States Identification

Our research are based on real life data acquired over 4 years by current measurement system implemented on main motor drive of industrial press line presented in Fig. 4

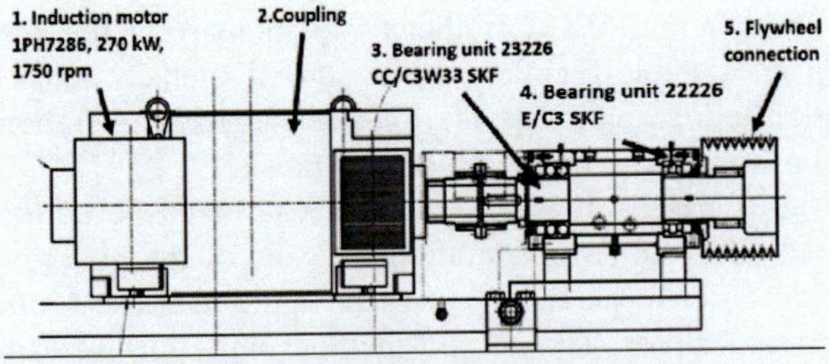

Fig. 4. Press main drive system

In Fig. 5 presents diagram of measurement system implemented on press main drive based on National Instrument Compact RIO modules. The purpose of the system is to perform electric current intensity and to acquire the data.

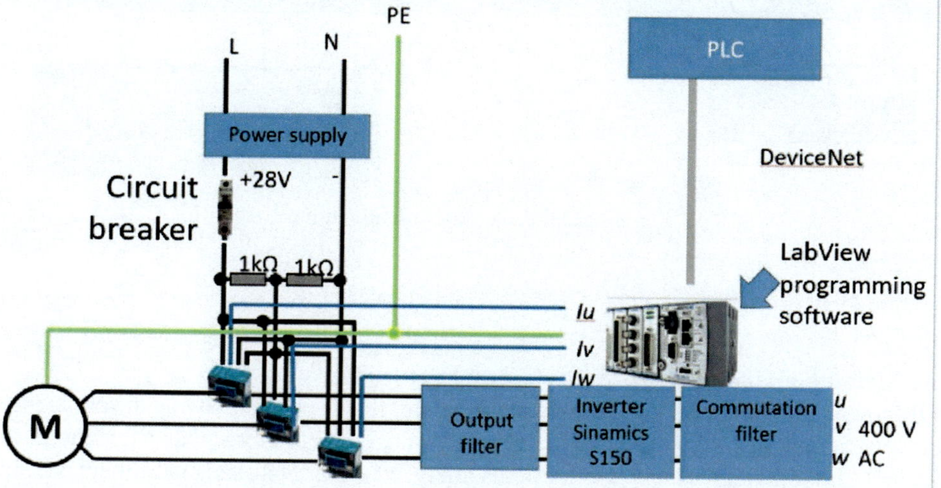

Fig. 5. Measurement system of press main drive system

The electric current is read with use of LEM 400-S current transducers with very high sampling frequency 100 kHz. Therefore, it is possible to analyze signals up to 50 kHz.

Over the time of signals acquisitions, it was possible to detect some problems with the monitored drive. It allows us to classify measured signals onto following categories Fig. 6:

State 0 - drive operating in normal conditions without any faults
State 1 - Broken bearings (4) between motor and flywheel connection and system misalignment
State 2 - Worn out elastic coupling, loosed electrical connection in the motor box, system misalignment
State 3 - Broken motor fan

The industrial reality presents that after two repairs the system response is different. It means that assembly of that size components size differs. That can be seen also on below picture. One may see easily that the hodographs differ. In order to design diagnostic system it is necessary to select characteristic feature of them that can be then analyzed.

In total 67 data sets have been collected:

- State 1: 2 data sets
- State 2: 21 data sets
- State 3: 19 data sets

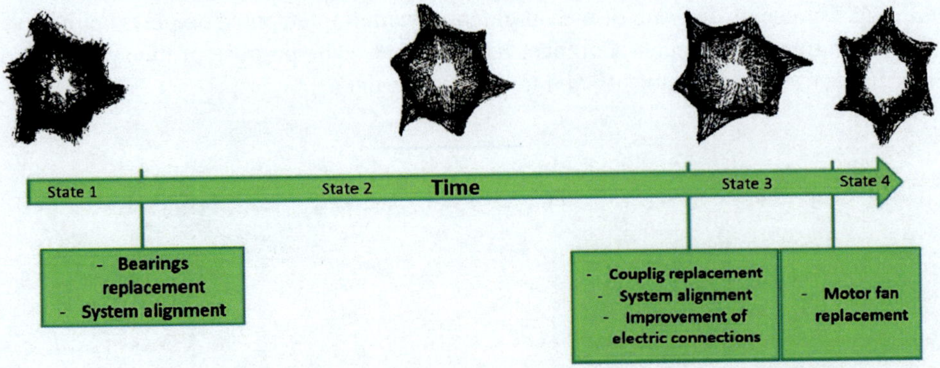

Fig. 6. Main drive states identification

- State 4: 24 data sets

The Method of Hodographs Analysis End Feature Extraction

Preliminary analysis of hodographs shapes indicates that is possible to find differences in geometrical features. It was motivation to apply image processing and analysis methods to extract hodograph features. Hodographs analyzed using National Instrument Vision Builder software, which allows us to perform the following sequence of image processing operations:

- Threshold
- Filter image by median filter
- Contour tracing
- Parameters definition
- Ellipse fit around Park's hodograph

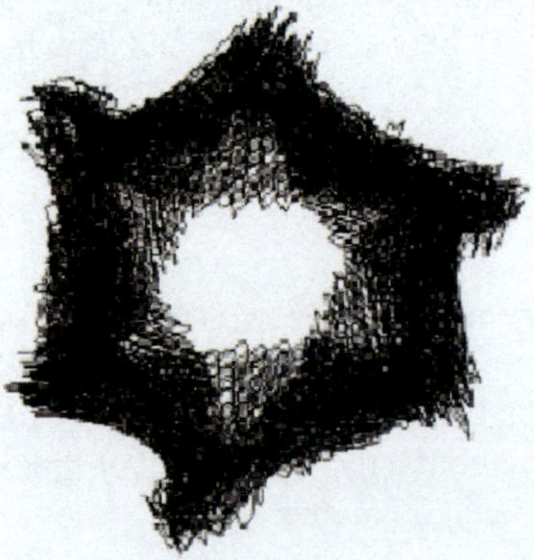

Fig. 7. Park's hodograph before vision processing

Hodographs before and after image processing are presented in Fig. 7 and 8. The main purpose of the vision processing was to close the hodograph and to find as much as possible features of its contour taking into account also the complete inner part of it. For that purpose it was necessary to utilize erosion and hole closure to have the solid shape. A separate analysis will be related also to the inner hole of the hodograph. Hodographs shapes were analyzed using blob analysis methods implemented in NI Vision Builder software. Almost 250 different topological image features were estimated for each processed hodograph image.

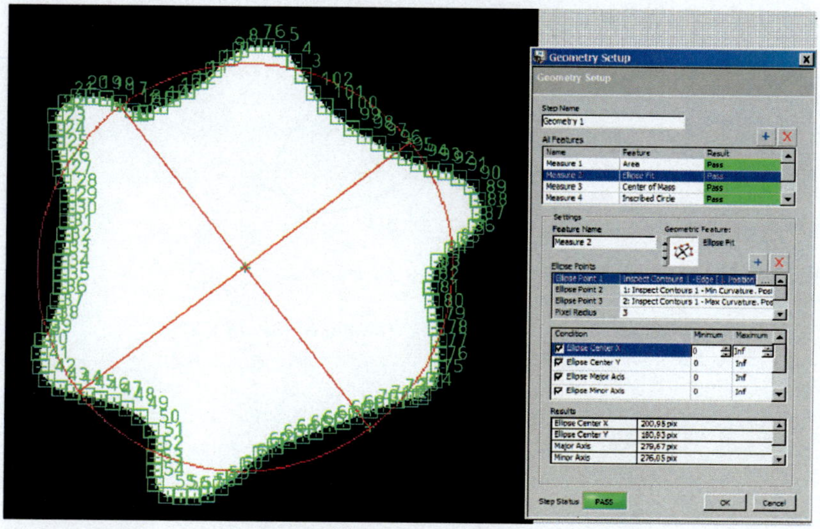

Fig. 8. Park's hodograph- vision processing

Figure 8 presents the vision processing of Park's hodograph. Applied filtering allowe to define 107 characteristic features that will be used for further analysis.

4 Characteristic Features

Some of a huge number of estimated hodograph features were redundant thus, it was necessary to limit the feature set to relevant one. The relevant features were selected on the basis of Pearson correlation matrix. Analysis of correlation coefficients allows us to indicate subset of 10 features for further considerations and conditions classification. The features are listed below.

- Object [1].Orientation (Pixels)
- Moment of Inertia XXY
- Moment of Inertia YYY
- Aspect Ratio
- Min Curvature.X Position (Pixel)
- Min Curvature.Y Position (Pixel)
- Max Curvature
- Max Curvature.X Position (Pixel)

- Max Curvature.Y Position (Pixel)
- Edge [1].Y Position (Pixel)

In Fig. 9 to 10 presented scatter plots of selected features values with indication of state classes. One can clearly see that is possible to discriminate groups of features values belonging to each condition class. It confirms that the features are suitable to perform classification process using one of commonly known minimal distance classifier like k-fold or SVM.

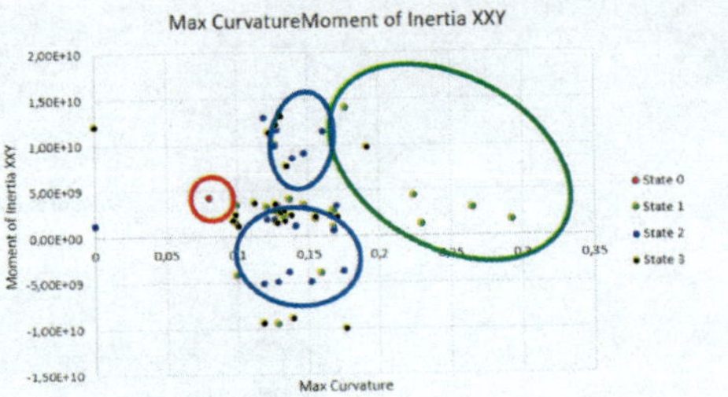

Fig. 9. Max curvature vs. Moment of inertia XXY

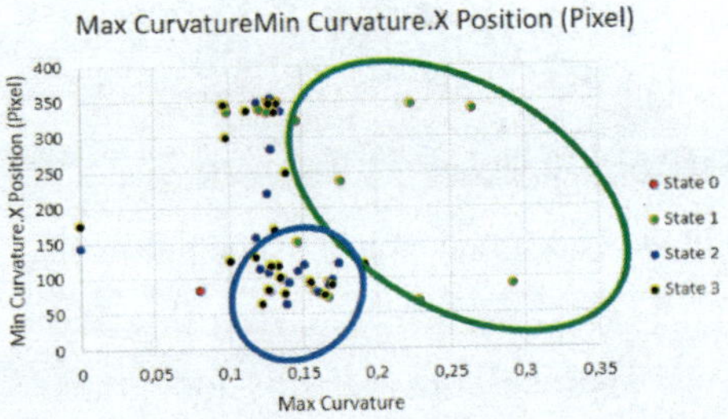

Fig. 10. Max curvature vs Min curvature X position

5 Conclusions

On the basis of results analysis one may come to the conclusions:

1. Due to limited amount of data related to malfunctions of the system it was necessary to combine different methods in order to achieve satisfactory analysis results
2. Lack of possibility to perform simulation of other malfunctions makes it impossible to predict if the method will identify the future data as malfunction
3. The method of Park's hodographs analysis gives satisfactory results when it is combined with correlation analysis of characteristic features. The next step of the research is too find effective classification algorithm which could be successfully implemented to industrial diagnostic system of press main drive.

References

1. Nia, N.K.: Efficient Simulation and Optimization for Tandem Press Lines. Thesis for the degree of Licentiate of Engineering at University West, University of Skövde School of Technology and Society (2012)
2. Roczek, K., Krol, A.: The influence of criticality analysis on methods selection strategy of press line monitoring system. In: 6th International Congress on Technical Diagnosis (2016)
3. Schuler Group, Smart Press Shop, Machine Monitoring System, 102ENG-102018-WSP-1000-Copyright by Schuler
4. SIEMENS, SINAMICS - Low Voltage Engineering Manual, SINAMICS G130, G150, S120 Chassis, S120 Cabinet Modules, S150, Version 6.5 July 2017
5. Ifm electronic, Automation Technology for the Machine Tools Industry 2014/2015
6. ABB, AbilityTM Smart Sensor. https://new.abb.com/motors-generators/service/advanced-services/smart-sensor
7. https://www.samotics.com/solution
8. https://www.koncarmonitoring.com/expert-motor-condition-monitoring.html
9. Bae, H., et al.: Fault diagnostic of induction motors for equipment reliability and health maintenance based upon Fourier and wavelet analysis. Artif. Life Robot. **9**, 112–116 (2005). https://doi.org/10.1007/s10015-004-0331-7
10. Castelli, M., Fossatti, J.P., Terra, J.I.: Fault diagnosis of induction motors based on FFT. In: Salih, S. (ed.) Fourier Transform - Signal Processing (2012). ISBN: 978-953-51-0453-7. http://www.intechopen.com/books/fourier-transform-signal-processing/fault-diagnosis-ofinduction-motors-based-on-fft
11. Kowalski, C., Ewert, P.: Zastosowanie analizy widmowej pradu stojan do monitorowania ekcentrycznosci silników indukcyjnych, Prace Naukowe Instytutu Maszyn, Napedów i Pomiarów Elektrycznych Nr 60 Politechniki Wrocławskiej (2007)
12. Mehla, N., Dahiya, R.: An approach of condition monitoring of induction motor using MCSA. Int. J. Syst. Appl. Eng. Dev. **1**(1), 13–17 (2007)
13. Penrose, H.: Motor Circuit Analysis Concept and Principle (2004). https://www.researchgate.net/publication/265193334_Motor_Circuit_Analysis_Concept_and_Principle, ALL-TEST Pro
14. Penrosem, H.W.: Applications for motor current signature analysis. An ALL-TEST Pro White Paper (2004)
15. Pillay, P., Xu Z.: Motor current signature analysis. IEEE (1996). 0-7803-3544-9/96$5.000
16. Roczek, K., Rogala, T.: Induction motor diagnosis wit use of electric parameters. ISSN 1641-6414, e-ISSN 2449-5220
17. Siemens, A.G.: Sinamics- Low Voltage Engineering Manual
18. Standard ISO 20958- Condition monitoring and diagnostics of machine systems—Electrical signature analysis of three-phase induction motors
19. Swędrowski L., Diagnostic measurements of current supplying an electric motor. In: Proceedings, XVII IMEKO World Congress, Dubrovnik, Croatia, 22 –27 June 2003, pp. 1426–1429 (2003)
20. Swędrowski, L.: Measuring system for analysis of motor supplying current for diagnostic purposes. In: SDEMPED 2005 - International Symposium on Diagnostics for Electric Machines, Power Electronics and Drives Vienna, Austria, 7–9 September 2005 (2005)
21. Thomson, W.T., Gilmore R.J.: Motor current signature analysis to detect faults in induction motor drives—fundamentals, data interpretation, and industrial case histories (2003). https://doi.org/10.21423/R1CS91
22. Parra, P.A., Palacios, J.A., Enciso, M.C.A., Olaya, J.: Stator fault diagnosis on squirrel cage induction motors by ESA and EPVA (2013). https://doi.org/10.1109/PEPQA.2013.6614937

23. Zarei, J., Hassani, H., Wei, Z., Karimi, H.R.: Broken rotor bars detection via Park's vector approach based on ANFIS, Conference Paper (2014)
24. Zarei, J.: Induction motors bearing fault detection using pattern recognition techniques. Expert Syst. Appl. **39**(1), 68–73 (2012). https://doi.org/10.1016/j.eswa.2011.06.042
25. Wolkiewicz, L., Kowalski, T.K.: Wykrywanie I lokalizacja zwarć zwojowych silnika indukcyjnego przy wykorzystaniu analizy składowych głównych hodografu wektora przestrzennego prądu stojana. Prace Naukowe Instytutu Maszyn. Napędów i Pomiarów Elektrycznych Politechniki Wrocławskiej **65**, 179–191 (2011)
26. Barnhart, B.L.: The Hilbert-Huang Transform: theory, applications, development (2011).https://ir.uiowa.edu/etd/2670/
27. Wang, L., Zhai, Y., Ji, S.: Realization of hilbert-huang transform based on LabVIEW and application in gearbox fault diagnosis. Boletín Técnico **55**(2), 10–19 (2017)

Switched Reluctance Motor Modelling Methods

Jakub Lorencki(✉) and Stanisław Radkowski

Warsaw University of Technology, Narbutta 84, Warsaw, Poland
Jakub.lorencki@pw.edu.pl

Abstract. A switched reluctance motor (SRM) can be a reliable electric drive operating in difficult environmental conditions. Despite the fact that this motor is inferior to some modern electric motors in terms of energy efficiency, it is recommended for vehicles drive for special applications: i.e. the military, mining, industry and other locations where reliability factor is more important. Since the motor core is made of iron, the magnetic characteristics are non-linear, therefore it is not possible to analytically model the operation of the reluctance motor. It is necessary to make the model using the finite element methods, here using the FEMM program. A 2D model was made and with the entire range of rotation (360 degrees) and for individual currents simulation and calculations of the magnetic quantities of the motor, such as the magnetic induction vector or inductance, were performed. As a result, 3D maps are created, which can be used later for dynamic modeling and another research.

Keywords: SRM · Motor diagnostics · FEMM

1 Introduction

1.1 SRM Overview

Switched reluctance motor (SRM) is an electric motor requiring position sensors and a voltage converter for power supply. The stator and rotor are made of iron and have characteristic salient poles. Only the stator is wound, rotor is passive. Although the principle of operation of these motors was invented at the end of the nineteenth century, it was not until the late eighties of the last century that they returned, began to come into more widespread use by the invention of semiconductors, power converters. The main study on SRMs is the monographs of Krishnan and Miller [1, 2], where the whole theory of these motors has been described. Ehsani et al. described the application of this motor in the context of a drive in alternative sources powered vehicles, such as batteries, fuel cells, hydrogen etc. [3]. Items [4–12] present the approach to SRM modelling under the simulation of dynamic eccentricity. The authors present the motor behaviour under the presence of misalignment.

Although these motors are not as energy efficient as e.g. BLDC permanent magnet motors, they possess other advantages. Due to the lack of magnets, they are resistant to demagnetization by high temperature or an electromagnetic pulse. In the event of an accident, such a magnet could also pose a certain risk.

A. Puchalski et al. (Eds.): ICTD 2022, ACM 21, pp. 105–114, 2023.
https://doi.org/10.1007/978-3-031-31719-4_11

The second advantage is the simplicity of its construction. Figure 1 shows its cross-section. The shape of this motor is quite simple to design and manufacture. For the above reasons, this motor can reach very high speeds, even exceeding 30,000 rpm. It can even function with a damaged phase.

Therefore, its application is rather designated for special purposes, such as military or industrial, i.e. those where reliability is more important, and a possible unexpected failure may cause irreparable losses in costs, and even in people.

This motor also has several disadvantages, most notably vibration, noise and torque pulsation. Another disadvantage is the demand for a voltage converter, motor control, and current and position sensors, although there are drive solutions that do not require position and current sensors.

1.2 SRM Design

As mentioned before, on the stator and rotor of the SRM there are salient poles and their number is different per motor. Typically, there are more poles on the stator than on the rotor, and the most common ratio of the number of stator poles to the rotor is 8/6, 6/4 or 12/8. The more poles, the less vibration, but the greater the cost.

Opposite poles form one phase and individual phases are supplied sequentially according to the rotation of the rotor. The asymmetric magnetic circuit associated with a different number of poles allows the generation of an electromagnetic torque. Certainly, in the coaxial position, when the axes of the stator and rotor poles coincide, the flux reluctance is the lowest, and vice versa, in the transverse position, when the poles are the farthest from each other - the highest.

1.3 Motor Voltage Equation

Generally, the voltage induced on the coil on the k-th phase is

$$U_k(i_k, \alpha) = \frac{d\varphi_k(i_k, \alpha)}{dt} \tag{1}$$

Taking into account the dependence of the flux associated with the k-th phase on m currents and the angular position, the value is

$$U_{ik}(i_k, a) = \sum_{i-1}^{m} \frac{\partial\varphi_k(i_k, \alpha)}{\partial i_l}\frac{di_l}{dt} + \frac{\partial\varphi_k(i_k, \alpha)}{\partial\alpha}\frac{d\alpha}{dt} \tag{2}$$

The first part of the Eq. 2 is transformation voltage, because they depend on the variability of individual currents over time. The second component is the rotation voltage as it depends on the angular speed of the motor. After the transformations, the general formula for the voltage in one phase of the motor is obtained

$$U_k(i_k, \alpha) = k_{Ek}(i_k, \alpha)\frac{d\alpha}{dt} + L_k(i_k, \alpha)\frac{di_l}{dt} + \mathrm{R}i_k \tag{3}$$

where k_{Ek} is the rotation induced voltage factor, R is the total phase resistance including the resistance of the transistor bridge.

2 SRM Modelling

2.1 FEA Analysis - Assumptions

An electric motor with the 8/6 configuration and the following dimensions was used for the tests (in mm) (Table 1):

Table 1. Motor Parameters

Stator Outer Radius	150
Stator Inner Radius	104.5
Stack Length	250
Height of Tooth	26
Width of Tooth Tip	26.1
Height of Tooth Foot	2.37
Width of Tooth	21.87
Air Gap	1
Rotor Outer Radius	103.53
Rotor Inner Radius	23.54
Width of Lobe	39.90
Height of Lobe	25.03

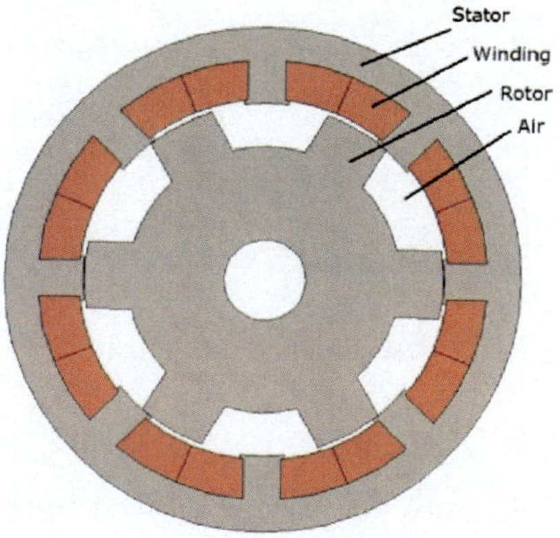

Fig. 1. Cross section of SRM of 8/6 poles

This motor geometry was introduced into the FEMM (Finite Element Method Magnetics) program to perform magnetostatic field calculations to obtain flux density B vector values. The calculations were made in 2D. This is a very common practice in such analyzes, because it can be assumed that the magnetic field does not change in terms of depth.

The measurement of the current density value was calculated at increment of one degree and all results are shown within one pole pitch of the rotor, i.e. 0–60°. The voltage across the coils ranged from zero to twenty incremented by four degrees. The loop and iterations were written in Matlab software, which cooperates with FEMM.

Figure 2 shows a fragment of the cross-section of the tested motor in the preprocessor mode with a generated mesh. You can see from this picture that the mesh is denser at the edges and thinner at larger surfaces. There are only three types of materials here: air, copper and steel (M-19).

2.2 Results

Figure 3 shows a three-dimensional diagram of the magnetic induction vector as a function of rotor angle position and current. With the coaxial position of the rotor and the stator, the function becomes saturated, while with the misalignment, the characteristic is linear.

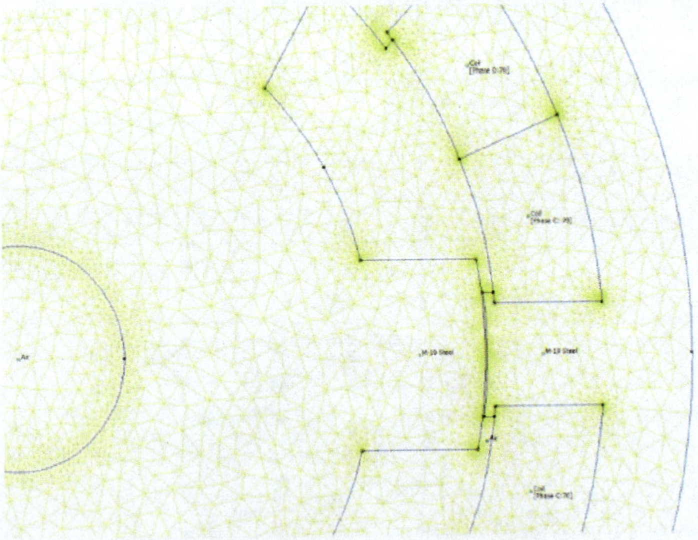

Fig. 2. Fragment of the machine cross-section and discretization into triangular elements of the machine

An important physical quantity that can be derived from flux density is dynamic inductance (Fig. 4). It is calculated from the formula

$$L_k(i_k, \alpha) = \frac{\partial \varphi_k(i_k, \alpha)}{\partial i_k} \tag{4}$$

At low currents, there are large differences between the transverse and coaxial inductance values. As the current increases, the values of the inductance in the coaxial position decrease, which also reduces the differences between the values of the inductance depending on the position of the rotor.

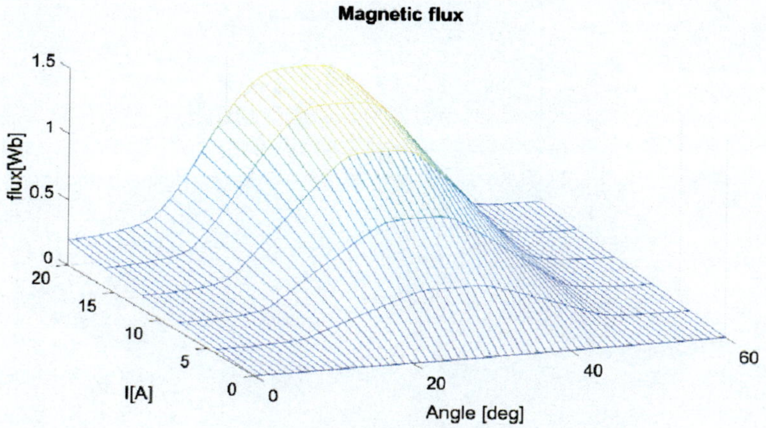

Fig. 3. Magnetic flux

Another calculated quantity is rotation induced voltage factor (Fig. 5)

$$k_{Ek}(i_k, \alpha) = \frac{\partial \varphi_k(i_k, \alpha)}{\partial \alpha} \tag{5}$$

In the transverse and coaxial positions, the coefficient is zero for any current value. The coefficient becomes positive when the rotor position is within the first half of the polar pitch, and negative when the rotor position is in the second half. The maximum values of the factor increase with increasing current.

Another value is magnetic coenergy (Fig. 6)

$$W_m^{'}(i, \alpha) = \int_0^i \left(i', \alpha\right) di' \tag{6}$$

In the transverse position, the magnetic coenergy is minimal similar to the flux density values.

The last determined value was the electromagnetic torque (Fig. 7)

$$\mathrm{T}_e(i, \alpha) = \frac{\partial}{\partial \alpha} W_m^{'}\left(i', \alpha\right) \tag{7}$$

The function is odd, zero-crossing in a coaxial position similar to the rotation voltage factor.

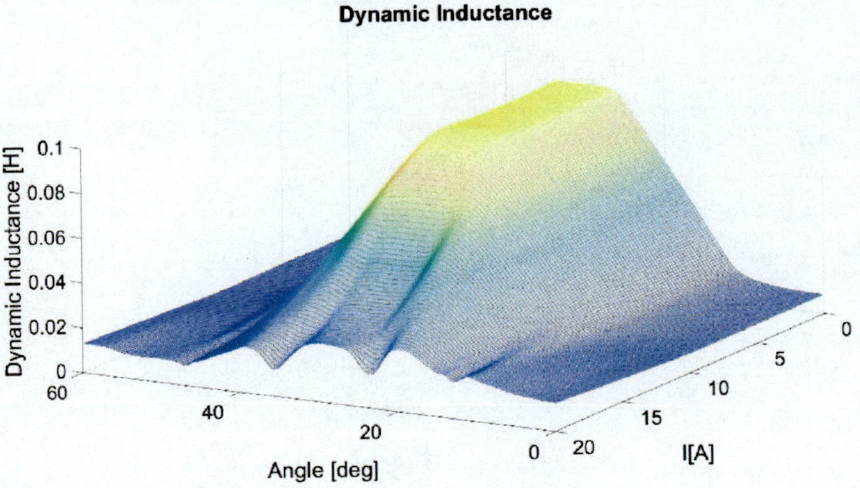

Fig. 4. Dynamic Inductance

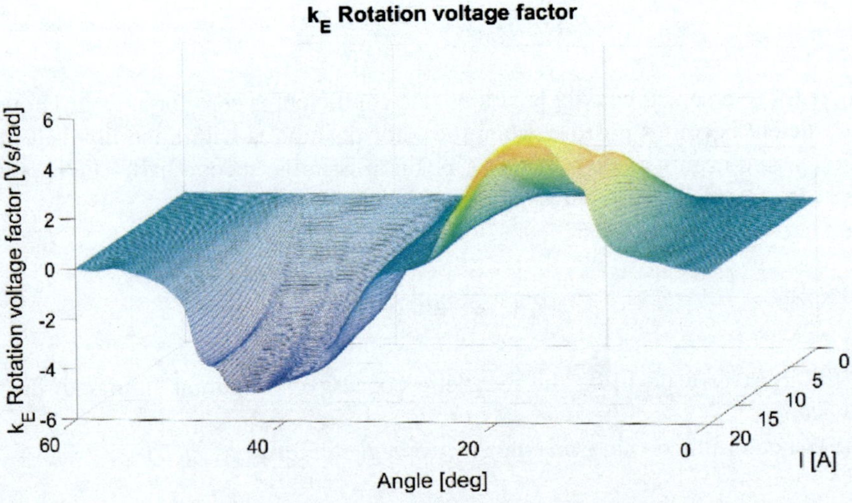

Fig. 5. Rotation voltage factor

Figures 8, 9 and 10 show two-dimensional plots of the magnetic induction vector, dynamic inductance, and the rotation voltage factor k_E at constant currents ranging from 4 to 20 amperes.

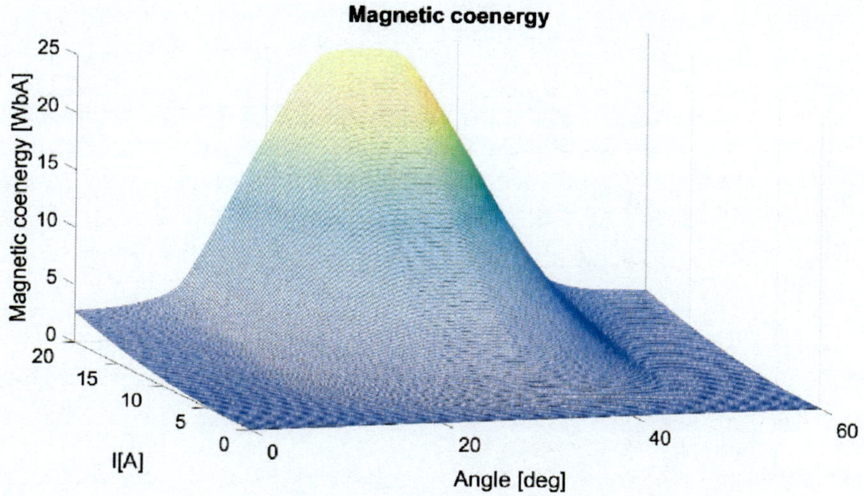

Fig. 6. Magnetic coenergy

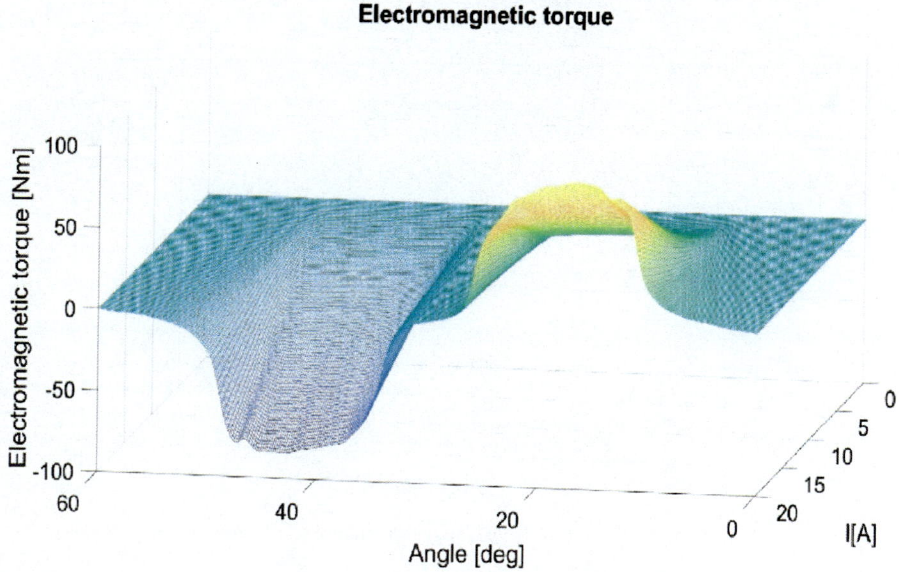

Fig. 7. Electromagnetic torque

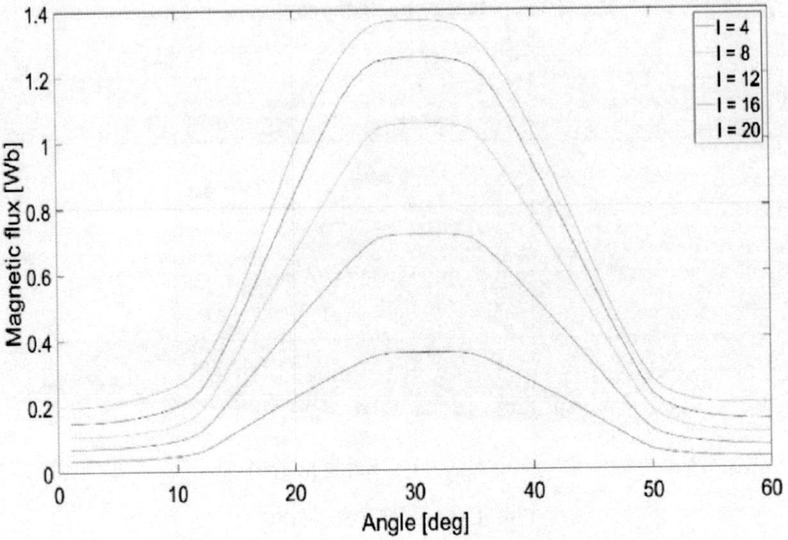

Fig. 8. Magnetic Flux for different currents

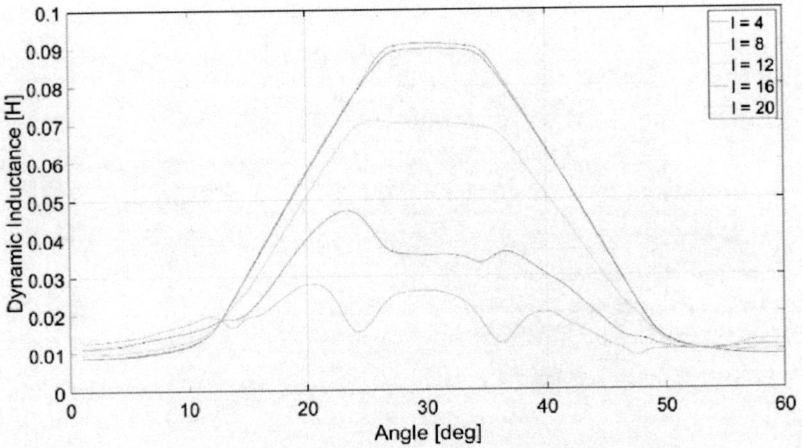

Fig. 9. Dynamic Inductance for different currents

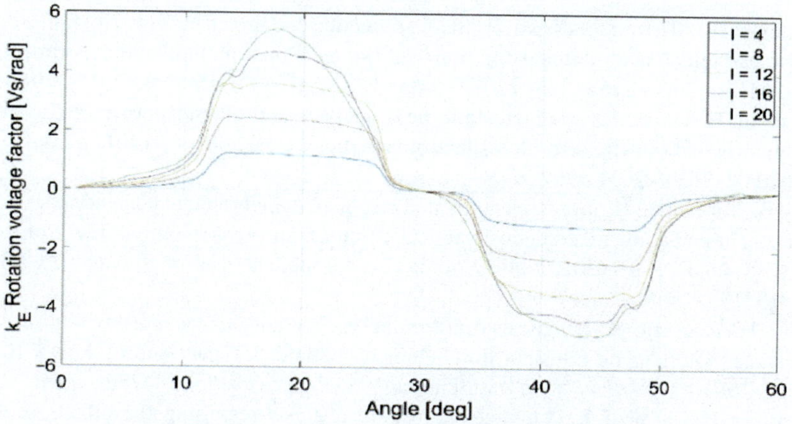

Fig. 10. Rotation voltage factor for different currents

3 Conclusions

In this work, we conducted a magnetostatic analysis in the environment of the finite element methods of the reluctance motor. We compared several physical quantities, such as the magnetic induction vector or dynamic inductance.

The research showed that despite the non-linear characteristics of the motor (related to the iron structure), the calculations show a large convergence with the physical parameters of the motor known from the literature.

Since this motor has nonlinear magnetization characteristics and therefore cannot be modeled by analytical equations, analysis in an FEMM environment or appropriate is the only necessary factor capable of initiating the motor simulation process for any other purpose. Magnetic data from this type of calculations are placed in an appropriate matrix (e.g. in the form known from loop-up tables from Matlab) and can be used for further analysis, e.g. to simulate dynamic states.

An analysis of this type may also lead to a legitimate introduction to further research.

References

1. Miller, T.J.E.: Switched Reluctance Motors and Their Control. Magna Physics Publishing and Clarendon Press, Oxford (1993)
2. Krishnan, R.: Switched Reluctance Motor Drives: Modeling, Simulation, Analysis, Design, and Applications. CRC Press, Boca Raton (2001)
3. Ehsani, M., Gao, Y., Longo, S., Ebrahimi, K.: Modern Electric, Hybrid Electric, and Fuel Cell Vehicles, 3rd edn.. CRC Press, Boca Raton (2018). https://doi.org/10.1201/9780429504884
4. Torkaman, H.: FEM analysis of angular misalignment fault in SRM magnetostatic characteristics. Progr. Electromagn. Res. **104**, 31–48 (2010). https://doi.org/10.2528/PIER10041406
5. Bouchareb, I., Bentounsi, A., Lebaroud, A.: Torque analysis of faulty SRM using FEA-TFR approach. J. Comput. Appl. **61**, 28–35 (2013). https://doi.org/10.5120/9916-4517

6. Torkaman, H., Afjei, E., Yadegari, P.: Static, dynamic, and mixed eccentricity faults diagnosis in switched reluctance motors using transient finite element method and experiments. IEEE Trans. Magn. **48**(8), 2254–2264 (2012). https://doi.org/10.1109/TMAG.2012.2191619
7. Torkaman, H., Afjei, E.: Magnetostatic field analysis and diagnosis of mixed eccentricity fault in switched reluctance motor. Electromagnetics **31**(5), 368–383 (2011). https://doi.org/10.1080/02726343.2011.579774
8. Afjei, E., Torkaman, H.: Finite element analysis of switched reluctance generator under fault condition oriented towards diagnosis of eccentricity fault. Appl. Comput. Electromagn. Soc. J. (ACES) **26**(1), 8–16 (2022). https://journals.riverpublishers.com/index.php/ACES/article/view/15213
9. Nazari, H., Rostami, N.: Diagnosis of different types of air-gap eccentricity fault in switched reluctance motors using transient finite element method. J. Oper. Autom. Power Eng. **3**(2), 94–101 (2015). https://www.sid.ir/en/journal/ViewPaper.aspx?id=504780
10. Torkaman, H., Afjei, S.E.: Magnetostatic field analysis regarding the effects of dynamic eccentricity in switched reluctance motor. Progr. Electromagn. Res. **8**, 163–180 (2009). https://doi.org/10.2528/PIERM09060205. http://www.jpier.org/PIERM/pier.php?paper=09060205
11. Torkaman, H., Afjei, E.: comprehensive magnetic field-based study on effects of static rotor eccentricity in switched reluctance motor parameters utilizing three-dimensional finite element. Electromagnetics **29**(5), 421–433 (2009). https://doi.org/10.1080/02726340902953354
12. Toliyat, H.A., Nandi, S., Choi, S., Meshgin-Kelk, H.: Electric Machines: Modeling, Condition Monitoring, and Fault Diagnosis, 1st edn. CRC Press, Boca Raton (2013). https://doi.org/10.1201/b13008

Diagnostics of Logistics Processes Using Pallet 4.0®

Tomasz Korbiel(✉)

AGH University of Science and Technology, Kraków, Poland
tkorbiel@agh.edu.pl

Abstract. The main processes in logistics are transport and storage. The introduction of standardized transport packaging has significantly improved these processes. The most popular transport packaging is a wooden EURO pallet. The industrial revolution 4.0 requires digital integration of elements of all processes. It was necessary to introduce innovative elements supporting Industry 4.0 to the transport pallet. A digital identification system, sensory systems as well as communication and power systems for the wooden transport pallet were developed. The digital identification systems are based on the UHF RFID standard. In the field of digital identification, the standard EPC number and a special user data space were used. Communication with sensory systems uses Bluetooth Low Energy. The acquisition system can optionally measure temperature, humidity, pallet tilt, vibrations and shocks, as well as the weight of goods on a pallet. The system uses solutions to reduce energy consumption through event definitions. Data acquisition takes place only in case of an event. This paper presents the assumptions for the construction of the system, a description of the developed systems and the results of tests of functional properties. The system was implemented for production and made available to a company that performs logistic processes.

Keywords: Logistics processes · process diagnostics · industry 4.0 · Transport pallet

1 Introduction

An important element of logistics processes in industry is transport and storage. The evolution of these processes significantly increased their reliability and efficiency, while reducing unit costs and losses [1]. The main factor behind this evolution was the introduction of transport packaging standards, in particular the pallet system. As a result, the following standards were introduced: pallet trucks, cargo spaces and storage systems. Their development and automation allowed for the introduction of an autonomous transport, high-bay warehouses, as well as advanced redistribution centers [2]. The foundation of these changes is a wooden pallet, invented over 60 years ago. In the era of Industry 4.0, the wooden pallet has also undergone a digital transformation [2]. It has been equipped with an RFID system, sensor systems, own digital identity as well as communication systems that fit into the IOT. The RFID system in connection with the databases identifies the indicated pallet, providing the necessary information. Digital identity allows you

A. Puchalski et al. (Eds.): ICTD 2022, ACM 21, pp. 115–122, 2023.
https://doi.org/10.1007/978-3-031-31719-4_12

to track the lifetime of the pallet, transport cycles, as well as protects against theft and fakes. The sensory systems continuously measure the environmental parameters: temperature, humidity, as well as mechanical influences and stability of the pallet, allowing the quality of logistic processes to be monitored. These and similar solutions contributed to the diagnosis of logistics processes, improvement of their quality, and the development of autonomous transport and storage systems. The obtained data allows for reliable management of logistics processes, in particular control, risk reduction and efficiency improvement. The article discusses in detail the solutions used in the Pallet 4.0®, presents an example of implementation and indicates the benefits of introducing diagnostics of logistics processes.

2 System Components

Standard goods packaging is the basic element of logistic transport and storage processes. For goods with a total weight of several-hundred kilograms, the collective packaging is a transport pallet [3]. For goods with a greater weight there are transport (sea) containers. The goods (cargo), after being placed on a pallet and secured against movement, constitute a compact body, easy to transport and store. Thus, the common element is a pallet, usually a wooden one [4]. This pallet has been equipped with electronic systems allowing for its identification, sensory systems and communication with other IT systems. Identification systems allow you to track the location of the pallet, while the measurement systems are used to identify transport and storage conditions.

2.1 Transport Pallet

The history of pallets begins with the issues of rail transport. The first pallet was produced around 1950 by the Swedish concern BT Industries [5]. From 1961, on the initiative of the UIC (International Railway Union), the pallets were standardized. After signing the contract, the pallet was given the status of standard and interchangeable. The Euro-pallet was designed and constructed mainly for the purpose of standardizing the size of pallets used in storage and transport. The dimensions of the euro pallet and its strength mean that the euro pallet allows you to optimally use the storage space, guaranteeing a low own weight and a static load capacity of up to 4,000 kg. Currently, the EUR (EPAL) pallet is one of the most used structures in all of Europe (Fig. 1) [6].

2.2 Radio Identification

The radio frequency identification system is based on the UHF RFID standard [7]. RFID (Radio Frequency Identification) is a technology that allows you to wirelessly identify almost any object by means of data transmitted over radio waves. There are three RFID standards that operate at low, high, or ultra-high frequency. The radio frequency ranges used for communication with the tags are: 125 kHz (LF), 13.56 MHz (HF) and 860–956 MHz (UHF). Each of these frequencies may be supported by different data transfer protocols. The UHF standard covers the frequency band from 300 MHz to 3 GHz. Due to this large discrepancy, the new UHF Gen2 standard was introduced, which uses the 860 to

Fig. 1. Standard EUR Pallet (EPAL)

960 MHz band and is supported by a single global protocol [8]. This protocol is adapted to European, Asian and American standards, and at the same time better optimized and more efficient. In the second generation of UHF, corrections were introduced that eliminated the errors occurring in the first generation by improving memory security and programming the Q algorithm to differentiate tags with the same serial number. In the UHF standard, the range of data reading and writing can be up to 20 m [9]. In the pallet, RFID tags are installed inside the wooden elements (Fig. 2).

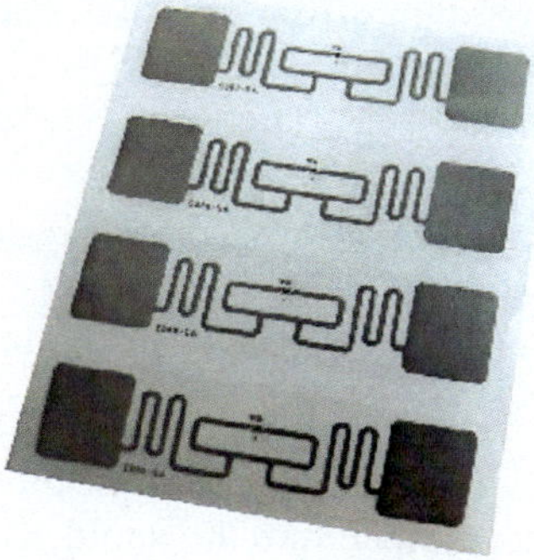

Fig. 2. Examples of UHF RFID tags

The programmable tag memory consists of an EPC (Electronic Product Code) number. 96bits, USER 512bits user memory, and one-time programmable TID 64bits memory. Tags are passive elements, their power comes from radio waves. The usage time is unlimited.

2.3 Measurement System

The lifetime of the pallet is several months up to several years. Thus, a system with a very low energy consumption was sought. The solution was the Bluetooth Low Energy (BLE) standard. The standard specification is supported from the Bluetooth 4.0 specification [10]. The measurement system is managed by the NRF51 or NRF52 series system by Nordic Semiconductror [11]. The block diagram of the system is shown below (Fig. 3).

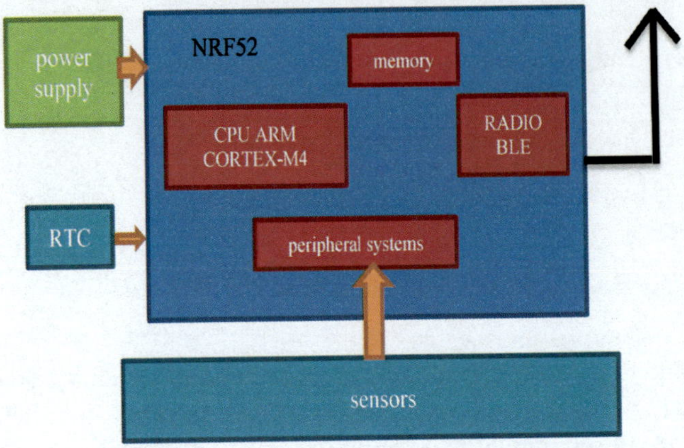

Fig. 3. Block diagram of the measuring system

The system enters the low power consumption (IDLE) state when the system is turned on. Only peripheral systems responsible for measuring physical values are supported. The system remains in this state until an event is detected. After detecting a predefined event, the system reads the time stamp from the RTC (real time clock) and starts the memory write procedures. The time stamp and the type of event with its characteristic parameter are saved in the memory [12]. This solution is characterized by low power consumption and saving space in data memory.

The system can be equipped with various sensors. Each of the sensors can be used optionally. In the prototype system, the sensors for temperature, humidity, weight and acceleration were installed. A number of events causing the start of memory recording have been defined, including:

- Change in temperature by more than 1.0 K since last measurement
- Exceeding the lower or upper temperature threshold
- Change in humidity by more than 5% since the last measurement
- Exceeding the upper humidity threshold
- Peak acceleration greater than 5 m/s^2
- Change of mean value of acceleration from the last measurement by more than 1 m/s^2
- Change in the mass of the goods by more than 2 kg since the last measurement
- Start BLE communication

The system was assembled on a miniature PCB circuit, containing power supply in the form of a CR2032 battery, PCB printed antennas and an integrated sensor in the form of an electronic circuit (Fig. 4).

Fig. 4. Assembled sensory system.

3 The Concept of a Global Pallet Identification System

Pallet communication systems (RFID-UHF, BLE) are standard solutions, allowing for integration with any ERP (Enterprise Resources Planning) management systems. A dedicated pallet identification and management system was also developed. The main idea of this system is total globalization. A generally accessible database will contain all information relating to each pallet. Access to data will take place at many levels of authorization, with the first public level. Inquiry at this level will allow the identification of the owner of the pallet, information about current certificates and other non-sensitive information. Higher authorization levels will allow to keep waybills, contact details of the sender and recipient, and administrative processes (customs, taxes). The block diagram of the system concept is presented below (Fig. 5).

In the figure, "Terminal" means any device that communicates with the pallet. The location of the terminal and its authorization status are known. After reading the data from the pallet, the terminal creates a message containing the information read, position data, and data status. This message is sent to the data server, where after verification, it constitutes an entry of the data record in the database. Data servers operate locally in individual companies or organizations. These servers exchange data with the global database system, where data is transferred from individual servers. Communication is two-way, data obtained from other servers belonging to other organizations via the database system are available for other servers. This information is also available to authorized customers for download. For example, the goods are sent from company A via the transport company B to the recipient C. Each of these companies has access to the current position of the goods (warehouse, transfer) as well as to the measurement data. This allows you to track the goods and the conditions in which it is transported. In the case of inspection by institution D, after prior authorization, it is possible to identify individual pallets in the warehouse or in the means of transport [13]. This solution is intended for the transport of sensitive goods (alcohol, hazardous materials, materials with limited availability). The data from the sensors, which are available through the database

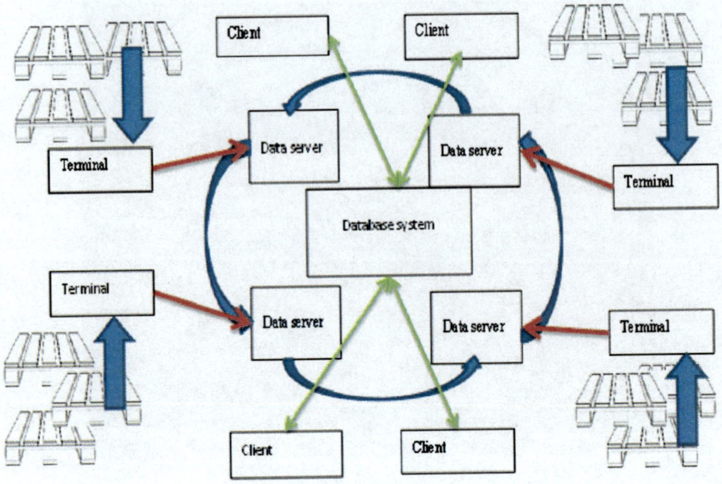

Fig. 5. Diagram of identification system concept

system, allows the company sending "A" and receiving "C" to possibly advertise the conditions of transport or storage in the event of poor quality of service.

4 Examples of the Use of the Pallet 4.0 in Industrial Conditions

4.1 Damaged Ceramics

A wholesale buyer of construction ceramics (roof tiles) received slightly damaged goods from the manufacturer. As the identification of the damage took place after some time, during the depalletization of the delivery, it was difficult to file a complaint. Pallet 4.0 was used for the selected transport. After receipt of the goods, the data was read and several individual shock pulses were indicated (Fig. 6). These were mechanical influences during transport, resulting in minor damage. The suspect was a freight transport company. However, the timestamp of the identified event indicated that the damage occurred during loading, not during transport. On this basis, an audit was carried out in the manufacturer's warehouse. The reason was a small threshold on the loading path and improper handling of the forklift. Too high speed of the trolley and careless handling of the goods resulted in its damage. After conducting the training and improving the conditions of internal transport, the problem was solved.

4.2 Transport of Food Products

Appropriate weather conditions are required in the transport of food products. Among other things, the temperature and humidity must be within the indicated ranges. With the use of Pallet 4.0, food products were transported to a large-format store. Receipt of goods has been refused due to failure to meet the transport conditions. After reading the data recorded in the memory of Pallet 4.0, it was found that the lower temperature threshold of transport was exceeded. After the investigation, it was found that the truck

Fig. 6. Example of damage to building ceramics (roof tiles)

driver had to stay overnight in the parking lot due to the time of his work. The low outside temperature at night caused a drop in temperature in the cargo space. The driver finished the course in the morning, but the conditions for transporting the goods were not met during the night. After finding any irregularities, the logistic processes were modified, inter alia, by the shipment of food goods in the morning or by the appropriate number of drivers, so that the goods reached the recipient immediately.

4.3 Shortage of Pallet Goods

In the transport between the warehouse and a large-area chemical store, shortages in the goods were noticed. The picked-up pallet with groupage goods was properly wrapped and had no signs of damage. The goods were picked up, but after some time, shortages were found in the unit packages. The 4.0 pallet with a weighing system was used to solve the problem. As anomalies during storage were suspected, pallet 4.0 was packed on the production line. From the production line, it was transported to a warehouse and then delivered to a retail store. Incorrect weight was found upon collection at the store. The weight change timestamps did not indicate any operations during storage. The weight change took place during the transport of goods between the warehouse and the store. The reason for the shortages was the driver who stopped during the transport after collecting the goods from the warehouse. During the stop, he removed one layer of goods from the pallet and then wrapped the rest of the pallet with the goods with foil. After the end of the procedure, the consequences were drawn and procedures were introduced to counteract a similar situation.

5 Summary

The transport pallet 50 years ago revolutionized logistics processes such as transport and storage. Since then, it has been produced in very large quantities unchanged. The industrial revolution 4.0 forced changes also in the transport pallet. The conducted research, design and implementation works ended with the introduction of a new transport pallet for use. It is equipped with digital identity, IOT-trend sensing systems and radio frequency identification systems. This design is in line with the ideas of Industry 4.0 and innovative trends in industrial logistics. The demonstration use of the presented solution made it possible to diagnose real problems occurring in transport and storage. This confirmed the great usefulness of the developed solution and the need to introduce this solution to mass production. The introduction of diagnostics of logistics processes to transport management is an investment with a short payback time [14]. Integration of the pallet 4.0 system with logistics management systems, such as ERP (Enterprise Resources Planning), will allow you to fully use the technical potential of Pallet 4.0 in the industrial revolution 4.0.

References

1. Ołdakowski, J., Ołdakowska, E.: Evaluation of unit costs of external transport effects used in cost-benefit analyses. Materiały Budowlane **1**(9), 105–107 (2022)
2. Korbiel, T., Pawlus, M., Gawronski, K.: Process of design and implementation of a digital transport pallet. In: 8Th Carpathian Logistics Congress (Clc 2018), pp. 191–196 (2019)
3. Kočí, V.: Comparisons of environmental impacts between wood and plastic transport pallets. Sci. Total Environ. **686**, 514–528 (2019)
4. Deviatkin, I., Khan, M., Ernst, E., Horttanainen, M.: Wooden and plastic pallets: a review of life cycle assessment (LCA) studies. Sustain **11**(20), 5750 (2019)
5. Martin, H.: Transport- und Lagerlogistik (2011)
6. Burdzik, R., Kabot, M., Cieśla, M.: Breakdown and internalization of external costs of road transport. In: Polish; Logistyka, no. 4 (2014)
7. Nikitin, P., Rao, K.V.S., Lazar, S.: An overview of near field UHF RFID. In: 2007 IEEE International Conference on RFID, pp. 167–174. IEEE RFID (2007)
8. Jin, L., Cheng, T.: Analysis and simulation of UHF RFID system. In: International Conference on Signal Process. Proceedings, ICSP, vol. 4 (2007)
9. Arnaud, A., Costa, G.: Ultra low-cost sensors using RFID standards for data collection, for IoT systems in food production and logistics. In: 2020 IEEE 11th Latin America Symposium Circuits System, LASCAS 2020 (2020)
10. Decuir, J., et al.: Bluetooth 4.0: low energy. Cambridge, UK Cambridge Silicon Radio SR plc **16**, 180 (2010)
11. Schrader, R., Ax, T., Röhrig, C., Fühner, C.: Advertising power consumption of bluetooth low energy systems. In: 2016 IEEE 3rd International Symposium Wireless System within IEEE International Conference on Intelligent Data Acquistion Advance Computer System, IDAACS-SWS 2016 - Proceedings, pp. 62–68 (2017)
12. Korbiel, T., et al.: Recognition of the 24-hour noise exposure of a human. Arch. Acoust. **42**(4), 601–607 (2017)
13. Yakimovich, B., Korshunov, A., Vladislav, S.: Increasing of the efficiency of flexible manufacturing system. Procedia Eng. **149**, 581–585 (2016)
14. Burdzik, R., Ciesla, M., Sładkowski, A.: Cargo Loading and unloading efficiency analysis in multimodal transport. Promet Traff. Transp. **26**(4), 323–331 (2014)

SYPIN: A System for Data Processing and Interpretation for Structural Health Monitoring

Ziemowit Dworakowski(✉)

AGH University of Science and Technology, Krakow, Poland
zdw@agh.edu.pl

Abstract. Recent years have seen a tremendous development of artificial-intelligence-based solutions. Although many areas of science and engineering were affected by this rapid growth, Structural Health Monitoring (SHM) appears to benefit from it to a relatively low extent. Despite the invention of methods that work with raw data, are general, and can be used to solve many complex challenges, most SHM approaches involve manual configuration of signal processing and decision-making routines. In order to develop universal solutions that do not require human input at every stage of processing, it is necessary to develop frameworks and methodologies that incorporate context knowledge, the ability to transfer information between systems and structures of varying levels of similarity, and the development of a robust and diverse database for decision systems training and evaluation. All of these requirements were addressed in the scope of the SYPIN project - aimed at developing an autonomous system that can replace human experts at every stage of SHM data processing.

Keywords: Structural Health Monitoring · Information processing · Recommendation · Artificial Intelligence · Autonomous decision making · Benchmark datasets

1 Structural Health Monitoring as a Data Science Problem

Artificial Intelligence (AI) obtained a prominent place in industrial data interpretation problems. Numerous examples can be shown to illustrate recent advances of artificial-intelligence-based solutions to solve large variety of difficult problems [1–6]. These approaches can be roughly divided into natural optimization, machine learning, and fuzzy-logic-based solutions [7]. However, even though enormous developments can recently be observed in AI-related fields, several challenges and practical issues undermine the practical availability of autonomous decision-making systems in industrial applications. Structural Health Monitoring (SHM) and Nondestructive testing (NDT) suffer from this issue in particular [8,9].

A. Puchalski et al. (Eds.): ICTD 2022, ACM 21, pp. 123–130, 2023.
https://doi.org/10.1007/978-3-031-31719-4_13

Any SHM decision system can only be as good as the dataset that it is built upon. Typical structure of SHM system is depicted in Fig. 1. In typical SHM cases the goal that is solved using the system is to detect, classify and assess future damage in given structure early enough, so the maintenance decisions can be made before the damage develops into a catastrophic failure. To meet this requirement, the system should first be given relevant data to learn to recognize general classes of undamaged and damaged systems. These data can come from three general sources:

Numerical simulation, where we design a numerical simulation of the monitored object with and without damages

Artificially introduced damage, where we use monitored object but artificially introduce *removable damage* - that is: simulated faults that can be introduced and deleted without affecting long-term structural health

Historical data from similar objects, where we assume that all objects from the same family share detectable reaction to early damage

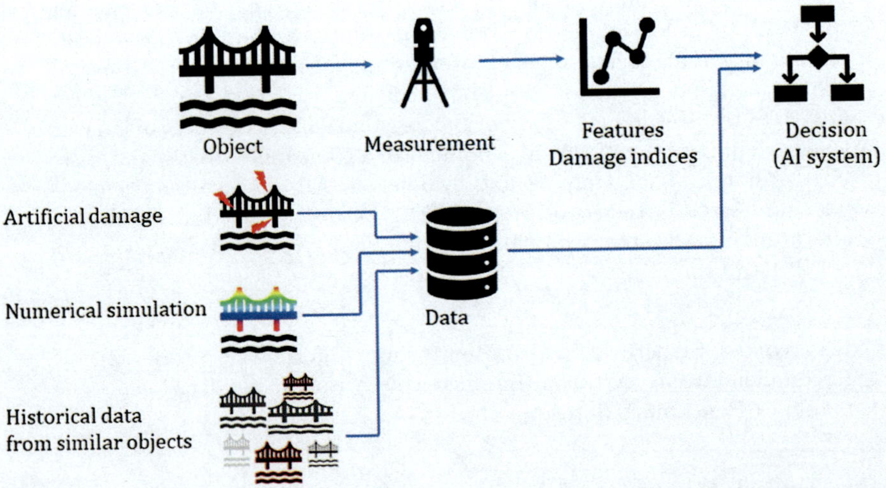

Fig. 1. Scheme of a typical SHM system with possible sources of training data

These three general approaches come with their own set of challenges in terms of providing enough similarity so that the generalization can be performedand, unfortunately, neither of them is usually efficient enough to be feasible in generation of large SHM datasets. Numerical simulations cost a lot of time and effort and, even if they can provide relatively diverse set of various damages and operational states, they tend to suffer from low similarity to real life scenarios. Artificially introduced damage allows to gather data on the same structure that will then be encompassed with monitoring, but artificial damage is rarely represetative to real-life wear and fatigue-induced faults. Similar objects, allowing for a population-based monitoring [10,11] are challenging in setup and require knowledge transfer, which is an ongoing research field. As a result, datasets used in SHM are often not diverse and general enough to be used in practice and should be perceived as more challenging than it is commonly understood.

2 SYPIN Project Principles

In order to build reliable SHM systems the following criterions should be fulfilled:

1. Ability to learn from a large variability of objects in the dataset
2. Ability to quickly process data for new objects - to enable easy scaling of a dataset
3. Ability to work in new conditions by transfer of knowledge from different objects from the past

These criterions led to design of a data processing and recommendation system within the scope of the SYPIN project [12]. The most important principle of the system is the capability to assess similarity of different objects in the database and recommend processing approaches to autonomously expand processed database.

Recommendation Module. The idea behind this section of the system is depicted in Fig. 2. Consider a new object for monitoring (1). Let us assume that we have a database of signals gathered on this object and that these signals contain information necessary for the detection of possible damage within the object (2). The task is to autonomously propose a processing and decision path that would lead to efficient damage detection (3). To this end we acquire context information about the object, that involves its type and important characteristics, conditions in which the object is monitored and the data that is currently available (4). Next, we calculate the similarity of each object within a historical database given the context (5), rank the objects in order of decreasing similarity and select several which are the most similar (6). For each of the selected objects we evaluate processing approaches that were used to solve that object's problems. Out of these approaches we select the one that appears to be the most efficient in historical setup (7). Finally, we calculate conditional probability of success given similarity between object under monitoring and selected object in database, and the relative efficiency of the approach in comparison to other approaches tested on the historical object. The processing chain with the highest likelihood is selected as the system's response (8).

Processing Module. The system is equipped with large variety of processing and classification solutions, including:

1. Domain change methods (including changes to frequency, time-frequency or order domain)
2. Filters (including frequency-based filtering and outlier removal)

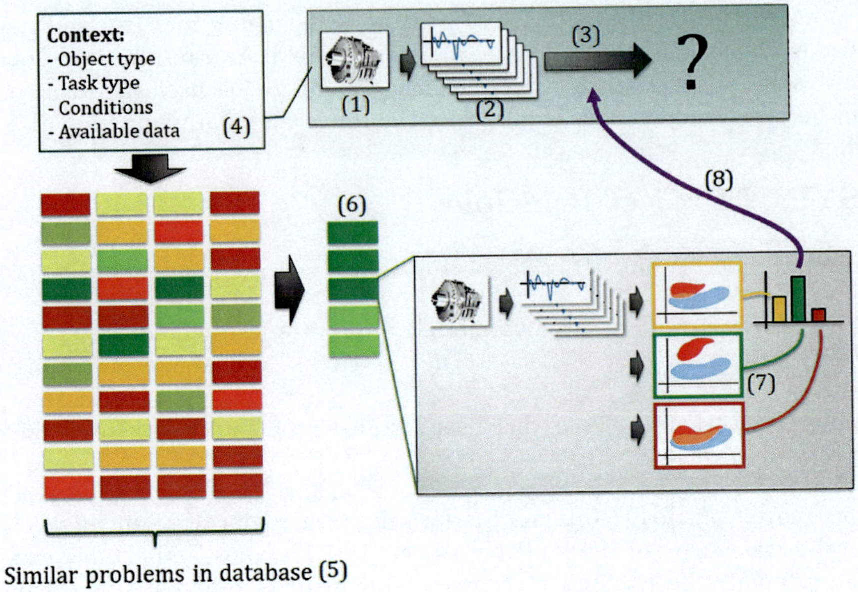

Fig. 2. Scheme of a recommendation module in SYPIN project. Objects are represented in the form of colored rectangles with color corresponding to the similarity between object in the database and object under monitoring.

3. Feature calculation (including statistical features, features specialized for condition monitoring and features specialized for guided wave signals)
4. Feature selection methods (including e.g. Principal Component Analysis and evolutionary-based approaches)
5. Novelty detection methods
6. Classification methods
7. Regression methods

Decision Module. The decision module of the SYPIN system works using a decision fusion approach. The information on many levels of processing is fused together time-wise and information-source-wise. The data in the database are stored in the form of *measurement series* where each series is treated as a snapshot of a structure's state in a given moment in time. Series may refer to several measurements taken over a short period of time using one or many different sensors. Information from each sensor can be processed independently and then combined using: majority voting setting - in which the final decision is made as an average of sub-decisions filtering out outliers, *whistleblower* setting - in which the final decision is based on the worst case within sub-decisions or *veto* setting in which the final decision is based on the best case within sub-decisions.

3 Benchmark Datasets

The data that were used to validate the proposed approach consisted of experimental results consisting of vibration and Guided Wave (GW) data. The latter set is made available to the general public in the form of a benchmark dataset, available at the project's webpage: **sypin.agh.edu.pl**. The GW dataset was built to enable experiments including knowledge transfer between specimens of different levels of similarity. There are three sections of the experiment corresponding to three sections of data: Aircraft specimens monitored during fatigue loading (Sect. A), Aircraft specimens monitored during variable environmental conditions (Sect. B), and composite specimens monitored for Barely Visible Impact Damage (BVID, Sect. C). All of the experimental data was acquired using the same DAQ and analogous setup, with the only difference being the sensors used for aluminum and composite specimens. The detailed setup of the experimental equipment is depicted in the Table 1. The data are in the form of measurement series which refer to the state of the structure at any given time. The details of each section are as follows:

Table 1. Acquisition system configuration

Parameter	Sections A and B	Section C
DAQ	EC Systems PAQ-16000D	EC Systems PAQ-16000D
Sensors	CTS Noliac NAC2002	Steminc SMD05T04R111WL
Sampling frequency	2,5 MHz	2,5 MHz
Extitation signal	250 Hz, 3 periods, Hanning-window-modulated	250 Hz, 3 periods, Hanning-window-modulated

Aircraft Specimens Fatigue Subset (Section A). Data in the first section contain results of fatigue experiments of three aircraft specimens. The specimens are depicted in Fig. 3. In each case the specimen was subjected to fatigue loading that caused development of a fatigue crack. The loading conditions were variable, the environmental conditions (temperature, in particular) were slightly variable. The total number of measurements within this section contain 202 measurement series.

Variable Environmental Conditions Subset (Section B). Data in the second section contain results of extended experiments performed under variable temperature. The specimens, similarily to section A, were aircraft wing components and are depicted in Fig. 4. Similarity between specimens allows for a possible knowledge transfer. Two of the specimens are covered in sealant since they are taken from fuel tank section of a wing. The temperature during the experiment was changed in a 0 to 67 °C range. The total number of measurement series was equal to 63.

Composite Specimens Subset (Section C). The third section of the research contain composite specimens with embedded sensors - depicted in Fig. 4. This subset was devoted to detection of Barely Visible Impact Damages (BVID) based on data acquired previously through introduction of artificial damage - in a form of attached masses. The BVID damage was introduced using an airsoft gun.

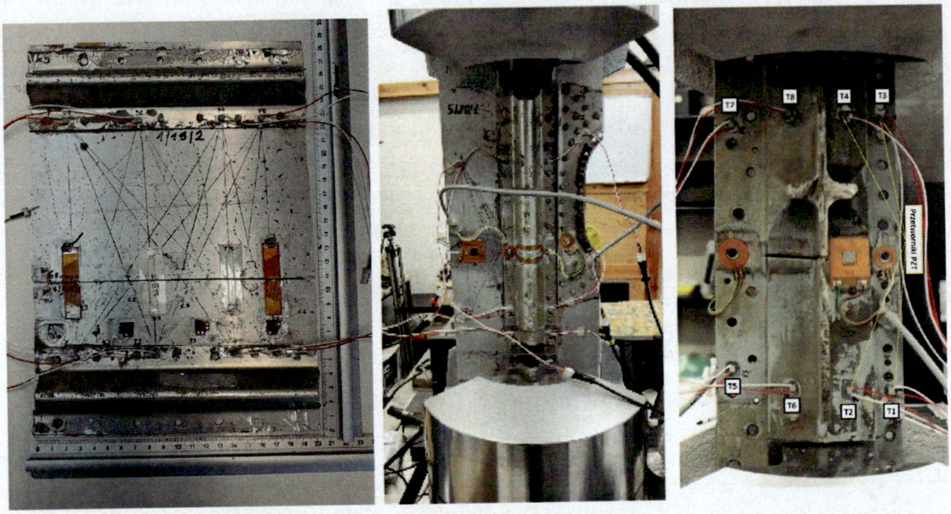

Fig. 3. Specimens used in section A of the experiment

Fig. 4. Specimens used in the section B of the experiment

Similarily to the Section B, here also the temperature was changes significantly during the experiment. The range of changes was -2 to 70 °C. The total number of measurement series among the three specimens was 81.

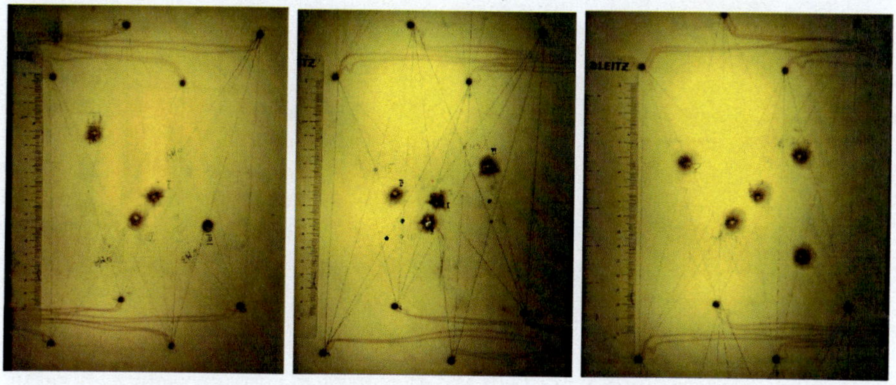

Fig. 5. Specimens used in the section C of the experiment

4 Summary and Conclusions

The SYPIN project resulted in a design of a decision system able to propose recommendations for processing and interpretation of data based on historical databases. Due to this capability it locates itself within other population-based approaches and provides a way to automatically solve new SHM tasks. The core of the system is a recommendation module that is able to assess the similarity between different objects in database in terms of their physical parameters, task type and dataset features. In order to test and evaluate the proposed methodology a large and diverse dataset was built, which is made available in the open-access form, as a benchmark dataset.

Acknowledgement. The work presented in this paper was supported by the National Center for Research and Development in Poland, under project number LIDER/3/0005/L-9/17/NCBR/2018

References

1. Längkvist, M., Karlsson, L., Loutfi, A.: A review of unsupervised feature learning and deep learning for time-series modeling. Pattern Recogn. Lett. **42**(1), 11–24 (2014)
2. Jablonski, A., Dworakowski, Z., Dziedziech, K., Chaari, F.: Vibration-based diagnostics of epicyclic gearboxes - From classical to soft-computing methods. Measurement **147**, 106811 (2019)
3. Kiumarsi, B., Vamvoudakis, K.G., Modares, H., Lewis, F.L.: Optimal and autonomous control using reinforcement learning: a survey. IEEE Trans. Neural Netw. Learn. Syst. **29**(6), 2042–2062 (2018)

4. Wu, R.-t., Jahanshahi, M.: Data fusion approaches for structural health monitoring and system identification: Past , present , and future. Structural Health Monitoring, Online first (2018)
5. Vázquez-Canteli, J.R., Nagy, Z.: Reinforcement learning for demand response: A review of algorithms and modeling techniques. Appli. Energy **235**(2018), 1072–1089 (2019)
6. Mahdavifar, S., Ghorbani, A.A.: Application of deep learning to cybersecurity: a survey. Neurocomputing **347**, 149–176 (2019)
7. Worden, K., Staszewski, W.J., Hensman, J.J.: Natural computing for mechanical systems research: A tutorial overview. Mech. Syst. Signal Process. **25**, 4–111 (2011)
8. Cawley, P.: Structural health monitoring?: Closing the gap between research and industrial deployment. Struct. Health Monit. **17**(5), 1225–1244 (2018)
9. Antoni, J.: A critical overview of the "Filterbank-feature-decision, methodology in machine condition monitoring. Acoustics Australia **49**(2), 177–184 (2021)
10. Bull, L.A., et al.: On the transfer of damage detectors between structures: An experimental case study. J. Sound Vib. **501**, 116072 (2021)
11. Shen, F., Langari, R., Yan, R.: Transfer between multiple machine plants: A modified fast self-organizing feature map and two-order selective ensemble based fault diagnosis strategy. Meas. J. Int. Meas. Confederation **151**, 107155 (2020)
12. Gorski, J., Heesch, M., Dziendzikowski, M., Dworakowski, Z.: Fuzzy-logic-based recommendation system for processing in condition monitoring. Sensors **22**, 3695 (2022)

Thermal Signature of Weld Defects Inside High Strength Steel Butt-Welded Joint Pipes

Massimiliano Pedot(✉) and Anna Timofiejczuk

Faculty of Mechanical Engineering, Silesian University of Technology, Konarskiego 18A, Gliwice, Poland

{mpedot,anna.timofiejczuk}@polsl.pl

Abstract. Design of complex infrastructures requires a multidisciplinary approach to prevent Natech accidents, which are secondary effects due to natural hazards. This approach follows two main directions: i) prevention of the loss of control over risk and ii) recovery from the loss of control over risk. These concepts may be applied to piping system installations, which are key elements of petrochemical and refinery plants. They are critical components that require special attention towards hazardous events because their rupture could cause leakage of the content resulting in possible explosions. Piping systems for oil and gas applications are generally an interconnection of carbon steel linear pipes, elbows, pipe fittings and auxiliary elements joined together using welding technology. The connection between the elements is a sensitive part that could trigger failure mechanisms in case of extreme loads. In fact, the manufacturing of the joints can take place under different condition e.g., technology adopted, manual or automatic welding process, manufacturing location, significantly affecting the final quality of the connection. It is important then to adopt proper diagnostic methods that allow to assess the development of imperfection inside the welded joints ensuring optimal quality acceptance levels. In this paper is described the possibility to adopt infrared cameras for this purpose. During preliminary experimental activity, were manufactured several pipe specimens using three different manual welding technologies. Each specimen consisted in two portions of X80 high strength steel pipe (diameter 12"$^{3/4}$) connected through butt-welded joints having induced imperfections inside. The actual position and the magnitude of these imperfections were then determined using radiographic analyses. This information was subsequently combined with the measurements of the welding pool temperature, which is the temperature of the molten material along the joint and recorded during the manufacturing process using an infrared camera. This process allowed to obtain the thermal signature of each specific defect, which is the temperature of the molten material that is associated to the development of the imperfection.

Keywords: High strength steel pipes · butt-welded joints · weld defects · monitoring system · petrochemical plants · welding pool temperature

A. Puchalski et al. (Eds.): ICTD 2022, ACM 21, pp. 131–143, 2023.
https://doi.org/10.1007/978-3-031-31719-4_14

1 Introduction

1.1 Background and Motivations

The effects on industrial plants or critical infrastructures resulting from natural hazards (e.g., earthquakes, floodings, impact crashes) are called "Natech" accidents [1]. Major examples of Natech accidents are the nuclear meltdown at Fukushima Dai-ichi nuclear power plant hit by a tsunami in 2011 [2] or hydrocarbon spills from several petrochemical plants in Louisiana caused by hurricane Katrina in 2005 [3]. The attempts to mitigate the effects of catastrophic events, can follow two different paths. The first one, highlighted in blue on the left side of Fig. 1, represent the measures that is possible to adopt to minimize the hazard. Examples of this measures can be refined hazard prediction models, innovative design of the elements to better withstand the threat, methods for rapid detection of faulty or damaged critical components. The second group of countermeasures, which are highlighted in green on the right side of Fig. 1, represents all the actions meant for the protection of the surrounding community and environment. In this sense, examples are the presence of specific barriers for the containment of the damage, safety mechanism used to shut off machinery in an emergency.

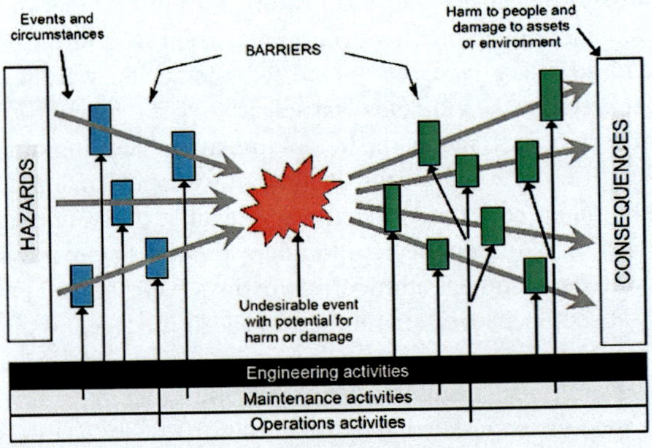

Fig. 1. Bowtie model [4]

Concentrating the attention on petrochemical plants, it is possible to identify various critical components. These elements are essentially storage tanks, piping systems, concrete or steel support structures, process units.

The design and material properties of piping systems are determined by functional requirements for the transmission of fluids from one point to another e.g., fluid transported, desired flux and velocity, space limitation, stress analysis. In general, the connection between all the elements is established by means of two types of joints: bolted flanged or welded (Fig. 2 and Fig. 3). Bolted flanged joints are non-permanent connections made coupling the plain faces of two flanges by mean of a variable number of bolts. In this case it is of particular importance to verify the health of the sealing gasket to prevent leakage of the content that could cause explosions. The second type of joint is instead a permanent connection between the ending faces of two pipes that are bonded together using an external source of heat that melts a portion of the material.

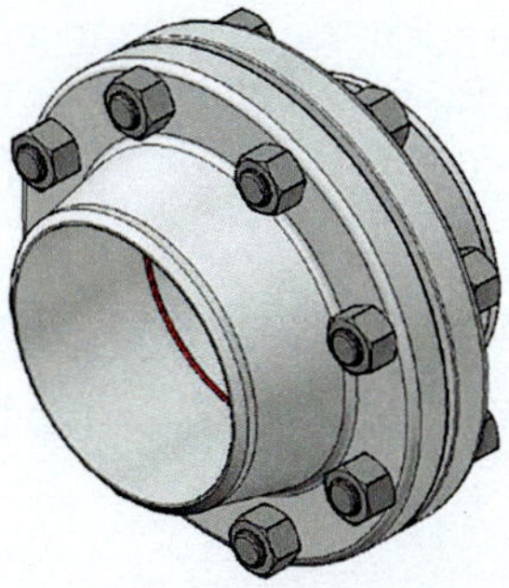

Fig. 2. Bolted flanged joint

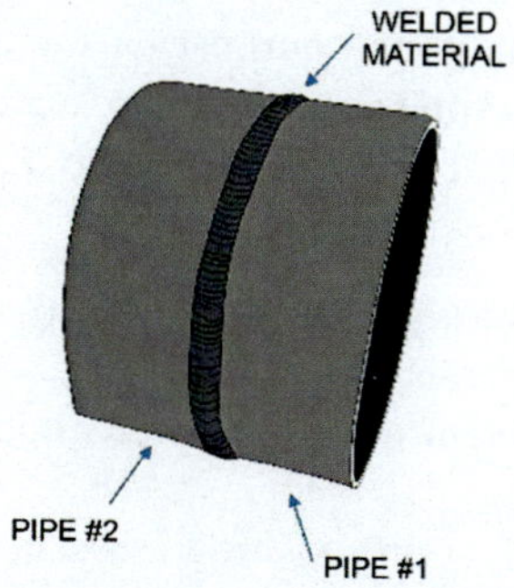

Fig. 3. Butt-welded joint

For this second class of joints, manual or semiautomatic welding process is a critical phase because external contaminations are more likely to occur, affecting the overall outcome with the growth of unwanted imperfections inside the weld metal. Common imperfections are e.g., porosity, lack of penetration, undercut, metallic inclusions, root concavity. The rupture of the welded joint due to ultimate limit state loads can be favoured by the presence of defects that could cause potentially dangerous stress concentrations. External factors that could favour the development of defects are e.g., presence of impurities (dust) or substances (water, grease) on the pipe surface, reduced flux of shielding gas, wrong arc voltage or welding current parameters selection.

1.2 Scope

In the previous paragraphs it was pointed out that Natech accidents prevention can be approached following two directions: mitigation of risk and protection from adverse consequences. Approaching the study of butt-welded joints for steel pipes installations, it was decided to concentrate the attention on the possibility to develop new techniques for rapid detection of weld imperfections. The focus was put high strength low alloy steel (HSLA) pipe class that represents a valid solution for oil and gas industrial applications given its increased strength property. The most common industry standard for steel line pipe is the API 5L [5]. This standard specification covers both normal (grade B with yield strength of 35,000 psi/240 MPa) and high-strength steels (X42 to X80 with yield strength of 42,000 psi/290 MPa to 80,000 psi/550 MPa) [6].

Given enhanced mechanical properties, it is possible to adopt pipes with reduced wall thickness. This implementation results in pipes weight reduction as well as the amount of welding consumables necessary for the manufacture of welded joints.

Research activities regarding high strength steels focus mainly on manufacturing technology and methods for the optimization of production processes. Examples are [7–9] that focus on material science aspects like chemical composition, cooling temperature and thermomechanical treatment process in order to enhance mechanical properties and satisfy quality acceptance levels. In this research was instead assessed the adoption of non-destructive method, based on thermal imaging technology, for quality assessment of butt-welded joints of HSLA steel pipes for typical petrochemical installations.

In details, the purpose of the experimental activity was the development of weld defects thermal signature map, which can be considered as the critical welding pool temperature that is correlated to the development of the single defects. The work consisted of the following steps:

- study of typical petrochemical plants and selection of one steel pipe with compatible diameter as case study
- manufacturing of butt-welded joint pipe segments with artificially induced imperfections
- imperfections position assessment
- analysis of infrared camera welding pool temperature records and definition of defects signature maps

In the second paragraph typical petrochemical plants, and in detail their piping systems, were analysed. Following this assessment, a 12"3/4 API-5L X80 pipe were selected and bought as suitable element for the experimental activity. In the same paragraph were also described the geometrical and mechanical properties of the pipe selected. In the third paragraph were illustrated all the process regarding the manufacturing of butt-welded joint pipes for the experimental activity. A total number of twelve samples were manufactured with induced imperfections inside the welded joints. On the following paragraph was introduced the process of identification of every single imperfection using radiographic method. For any of them it was assessed the exact position and its magnitude. In the last paragraph is described the comparison between the information obtained from radiographic analysis and the data regarding welding pool temperature recorded during the manufacturing process using an infrared camera.

2 Case Study Pipe Selection

2.1 Typical Piping System Configurations

Before starting the manufacture of butt-welded joint pipe specimens, it was necessary to perform an assessment of typical petrochemical plants layout. This was essential for the correct selection of a pipe with comparable dimensions for the subsequent experimental activity. Typical configurations of piping systems installed on petrochemical plants consist of a combination of different elements (Fig. 4):

- linear pipe elements
- elbow elements connecting linear pipes placed on different levels or directions
- tee-joint elements that allow to diverge the flux into two directions
- reducer elements that connect two elements having different section diameter

Butt-welded joints connect two sections of linear pipes or pipe elements of various sizes laying on different levels and planes. Complex arrangements impose different welding positions, i.e., vertical and horizontal, during manufacturing process. In [11] is presented a seismic analysis of an actual LNG plant with complete piping system. The piping system consists of eight main pipes with different diameters connecting the storage tank to process areas of the plant. The properties of these eight pipes are listed in Table 1: pipes diameters are in a range from 4" (114,3 mm) to 18" (457,2 mm). It was

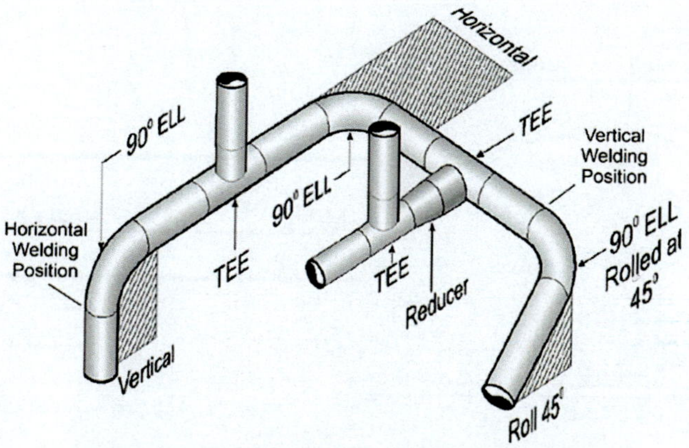

Fig. 4. Generic piping system configuration [10]

decided then to concentrate the experimental activity for this research on a pipe with comparable dimensions of those usually installed on real case plants (Fig. 6).

Table 1. Piping system installed on case study of [11]

Pipe	Pipe specification	External diameter [mm]	Wall thickness [mm]
1	16"- SCH20	406.4	7.92
2	10" - SCH10S	273.05	4.19
3	4" - SCH10S	114.3	3.05
4	6" - SCH10S	168.28	3.40
5	12" - SCH10S	323.85	4.57
6	6" - SCH10S 8" – SCH10S	168.28 219.08	3.40 3.76
7	6" - SCH10S	168.28	3.40
8	18" – SCH10S	457.2	4.78

2.2 Pipe Selected for the Experimental Activity

The first step of the experimental activity was the selection of a steel pipe with consisted steel grade and dimensions. For this research it was decided to put the focus on high strength steel pipes and for this reason it was considered API-5L X80 steel grade. Significant mechanical properties of the material are minimum yield strength $R_{t\,0,5}$ and ultimate tensile strength R_m; their values provided by the pipe manufacturer are reported in Fig. 5

The geometrical properties of the pipe were selected in accordance with the information collected in 2.1 to be comparable with real case piping systems dimensions. In

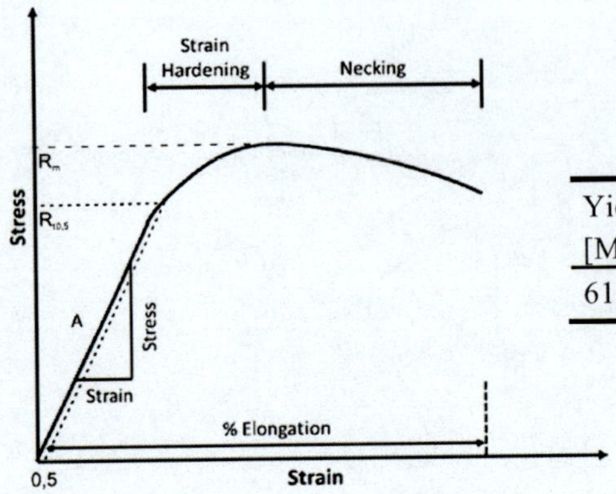

Yield strength $R_{t\,0,5}$ [MPa]	Ultimate tensile strength R_m [MPa]
615 ÷ 634	679 ÷ 699

Fig. 5. Mechanical properties of X80 grade steel [12]

detail it was chosen a 12" 3/4 diameter (323.9 mm) and 10 mm wall thickness hot rolled pipe.

3 Manufacturing of Butt-Welded Joint Pipe Specimens

3.1 Preliminary Activities

To perform the experiments, the pipe was cut in segments of about 20–25 cm of length. These pipe sections were then machined and clamped together as depicted in Fig. 6. A total number of twelve specimens was prepared for the subsequent welding procedure.

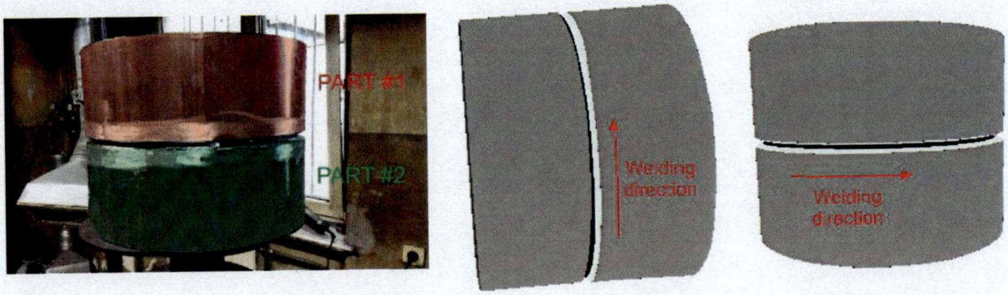

Fig. 6. Pipe manufacturing setup

This amount of pipe specimens was necessary to replicate various configurations regarding welding technology i.e., Gas Metal Arc Welding technology (GMA), Manual Metal Arc Welding technology (MMA) and Self Shielded Flux Core Wire Technology (SSA), welding position i.e., horizontal (PC) and vertical (PH) and position of the desired defects along the circumference. For each welding technology were manufactured four pipe specimens. Among each group of four, two samples were prepared adopting vertical welding position while the other two adopting horizontal welding position. The complete resume of the manufacturing plan is reported in Fig. 7.

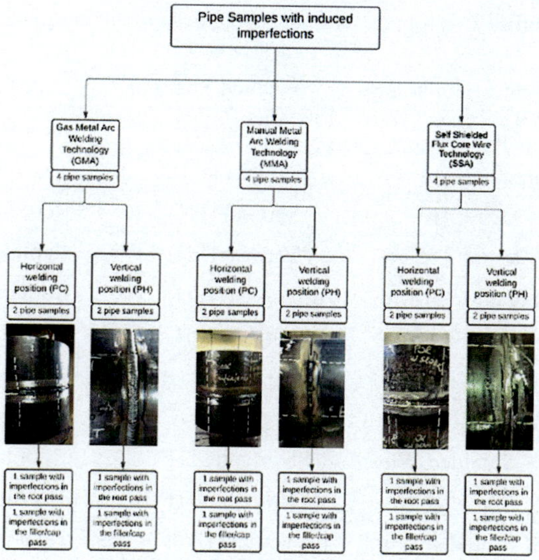

Fig. 7. Pipe specimens manufacturing plan

Before the welding process, the circumference of each pipe was divided approximatively into 8–9 different segments. This activity was meant to create defined portions where it was possible to apply controlled conditions on designated weld beads i.e., root bead and filler/groove bead. Examples of controlled weld conditions used to facilitate the development of the defects were the presence of water or grease on the pipe surface, intermittent shielding gas flux (for GMA welding technology), arc voltage and welding current parameters different from those prescribed by welding procedure specifications (WPS), different bead deposition speed. On Table 2 is reported the exact properties of each pipe specimen manufactured.

During the assembly process of all the twelve pipe sections, welding pool temperature of each segment was recorded using an infrared camera. The infrared camera had a resolution of 640 x 480 pixels and a sampling frequency of 50 Hz.

4 Quality Assessment of Butt-Welded Joints

The twelve pipe specimens were subjected to quality assessment process using radiographic technique with the purpose to localize and qualify the imperfections inside the butt-welded joints. The radiographic tests were carried out according to the standard ISO 17636–1:2013 [13] and were performed along the total length of every single joint. The imperfections detected were classified according to the standard EN ISO 6520–1 [13]. This activity allowed to create a complete report regarding defects characterization and their exact position along the circumference. The outcomes obtained from the analysis of the GMA3 pipe specimens are presented in this paper (Fig. 8–Fig. 9.)

The radiographic analysis revealed a great number of defects of different type inside the welded joint of GMA3 pipe. The same circumstance was observed on the other pipe specimens.

Table 2. Properties of pipe specimens manufactured

Specimen ID	Welding Technology	Welding Position	Imperfections Position
GMA1	Gas Metal Arc	Horizontal (PC)	Root pass
GMA2	Gas Metal Arc	Horizontal (PC)	Cap/filler pass
GMA3	Gas Metal Arc	Vertical (PH)	Root pass
GMA4	Gas Metal Arc	Vertical (PH)	Cap/filler pass
MMA1	Manual Metal Arc	Vertical (PH)	Root pass
MMA2	Manual Metal Arc	Vertical (PH)	Cap/filler pass
MMA3	Manual Metal Arc	Horizontal (PC)	Root pass
MMA4	Manual Metal Arc	Horizontal (PC)	Cap/filler pass
SSA1	Self-Shielded Wire	Horizontal (PC)	Root pass
SSA2	Self-Shielded Wire	Horizontal (PC)	Cap/filler pass
SSA3	Self-Shielded Wire	Vertical (PH)	Root pass
SSA4	Self-Shielded Wire	Vertical (PH)	Cap/filler pass

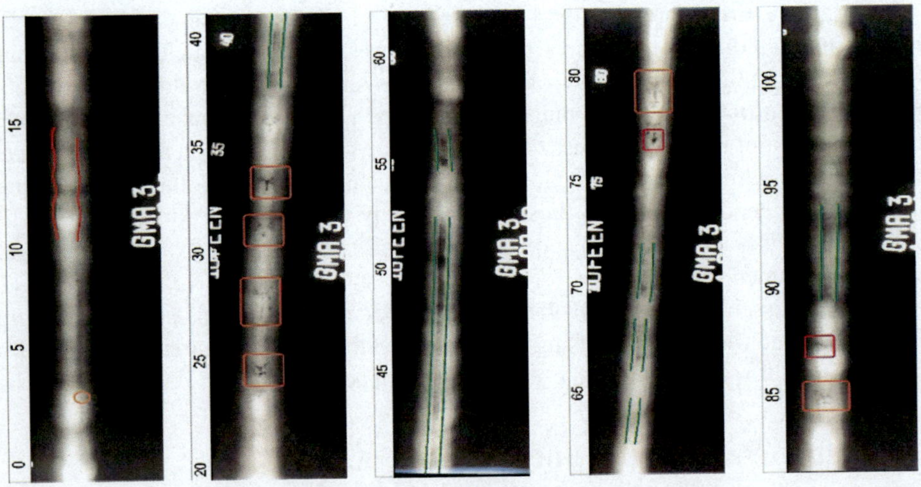

Fig. 8. Radiographic analysis of GMA3 pipe specimen

5 Thermal Signature of Defects Detected

The results obtained from the radiographic analysis on all the pipe specimens were combined with the welding pool temperature data collected with infrared camera during manufacturing process. The study of the temperature measured on the specific areas corresponding to the presence of single defects allowed to identify the thermal conditions that favour the development of those defects.

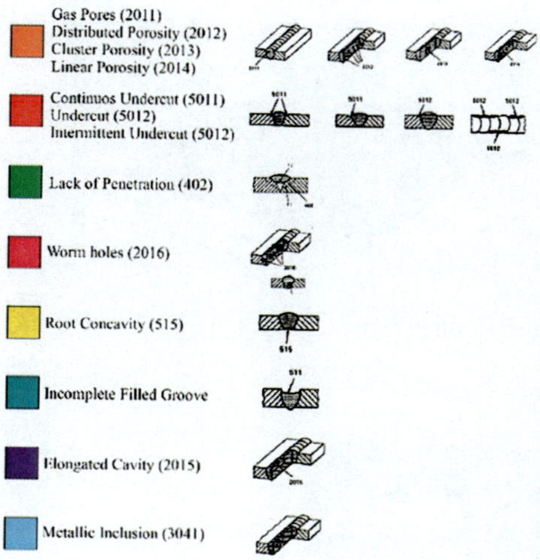

Defect detected	Position [cm]
Intermittent Undercut (5012) + Uniformly Distributed Porosity (2012)	3
Intermittent Undercut (5012) + Uniformly Distributed Porosity (2012)	9
Root Concavity (515)	14
Distributed Porosity (2012) + Worm holes (2016)	35
Distributed Porosity (2012) + Worm holes (2016)	44
Incomplete Filled Groove (511) + Gas Pores (2011)	51
Lack of Penetration (402) + Gas Pores (2011)	72
Lack of Penetration (402) + Gas Pores (2011)	78
Cluster Porosity (2013)	93

Fig. 9. Position of significant defects on GMA3 pipe specimen

In the next two paragraphs are presented the assessments carried out on two separate portions of the pipe marked as GMA3. The first case study regards the presence of cluster porosity inside the weld metal while the second case refers to lack of penetration defect.

5.1 Cluster Porosity Thermal Signature

On the pipe specimen GMA3, manufactured using gas metal arc welding technology and vertical welding position, it was detected cluster porosity on the portion of the joint between 25 and 35 cm along the circumference (Fig. 8.). The correspondent measure of welding pool temperature on this segment is depicted on Fig. 10. In the picture is represented the maximum temperature recorded at each point during the welding process of the root pass. From the analysis of these data, it was possible to notice the presence of five different pikes on the profile temperature.

Comparing the position of these peaks with the position of the porosity spotted using radiographic technique it is possible to observe complete correspondence between them (Fig. 11).

5.2 Lack of Penetration Thermal Signature

On the same pipe sample, at a distance between 40 and 55 cm along the circumference, was detected lack of penetration (Fig. 8). On this segment, the weld molten metal didn't

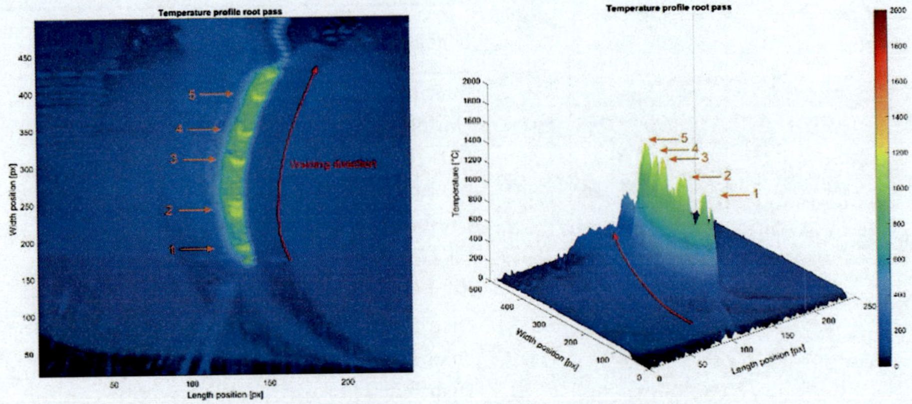

Fig. 10. Welding pool temperature for GMA3 pipe specimen between 24 and 38 cm along the circumference of the joint

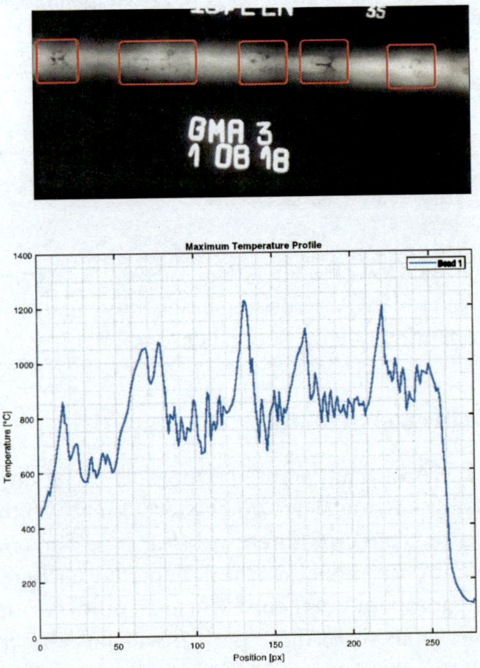

Fig. 11. Correspondence between welding pool temperature profile and defects position

fill completely the gap between the two pipes in the root area. The maximum values for welding pool temperature of each point are reported in Fig. 12.

In this case, a comparison was carried out between the temperatures recorded on this segment and those collected for the segment comprised between 15 and 25 cm where lack of penetration wasn't detected. The temperature profile of these two segments is

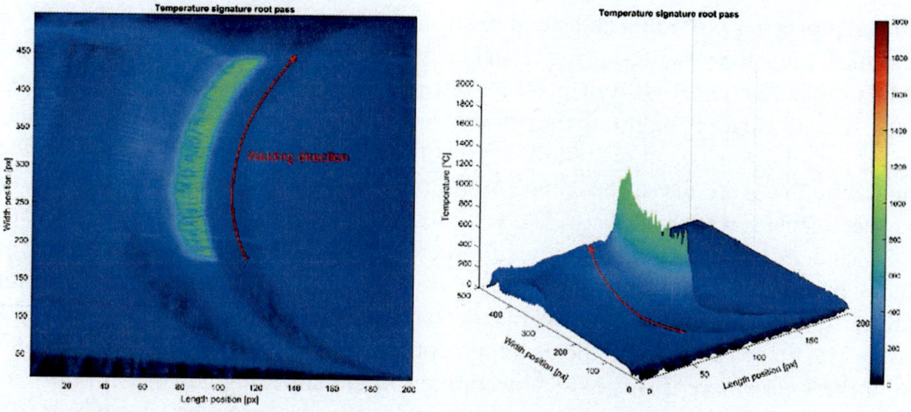

Fig. 12. Welding pool temperature for GMA3 pipe specimen between 40 and 55 cm along the circumference of the joint

reported in Fig. 13 and it is possible to observe lower values for the temperature where the defect is present.

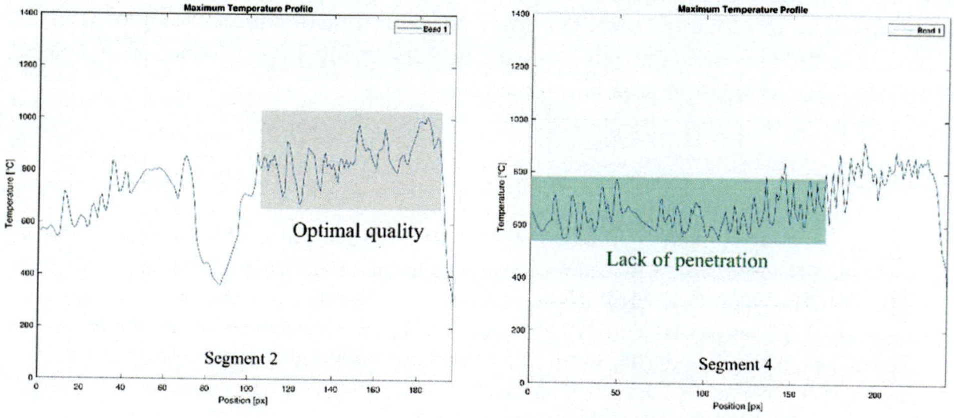

Fig. 13. Comparison between segment #2 and segment #4

6 Conclusions

The experimental activity performed consisted of the realization of twelve pipe specimens with induced imperfection inside butt-welded joints connecting two distinct sections. The research focused on high strength steel pipes which, given their enhanced mechanical properties, represent a interesting development in petrochemical plants applications. Weld imperfections were created simulating adverse conditions that could happen on real case scenarios during the welding process. The specimens were manufactured adopting different configuration regarding welding technology, welding position and bead properties providing at the end many cases to investigate. In fact, the quality of the welded joints was assessed using visual inspection as well as radiographic technique and this procedure allowed to identify 96 different portions that included at least one defect. The information regarding the position and the type of defect there were obtained with the quality assessment procedure were combined with the values of welding pool temperature that were measured using an infrared camera during the preparation of the pipe specimens. It was then investigated the possibility to identify a correlation, which can be named as defect thermal signature, between the presence of a specific defect with the thermal conditions in the welding area. In this paper were presented the results obtained from the analysis of two separate defects found on the pipe specimen marked as GMA3. The first case regarded the development of five clusters of porosity on a defined area of the joint where it was experienced intermittent shielding gas flux. In this case it was observed the presence of a peak of temperature on the same spot where each cluster was detected. The second example regarded instead lack of penetration of the melted metal on the gap between the pipes. In this case it was noted that the maximum temperature of the welding pool was lower than the temperature measured where the quality of the joint was optimal.

This experience suggests the possibility to replicate in the future the same process for the other segments where defects were detected. The analysis might provide further sets of defects thermal signatures for high strength steel pipes with various manufacturing conditions. The information obtained could then be used for the development of an automatic method for rapid detection of weld defects during joints production and based on the measures of welding pool temperature.

References

1. Krausmann, E., Cruz, A.M., Salzano, E.: Natech Risk Assessment and Management Reducing the Risk of Natural-Hazard Impact on Hazardous Installations. Elsevier (2017)
2. Eisler, R.: The Fukushima 2011 disaster. CRC Press - Taylor & Francis Group (2013)
3. Steinberg, L.J., Sengul, H., Cruz, A.M.: Natech risk and management: an assessment of the state of the art. Nat Hazards **46**, 143–152 (2008). https://doi.org/10.1007/s11069-007-9205-3
4. Hollnagel, E., Woods D, Leveson N (2006) Resilience Engineering : Concepts and Precepts
5. (2018) API 5L, Seamless & welded pipe
6. Stewart, M.: Pipeline systems. In: Surface Production Operations. Elsevier, pp. 813–884 (2016)
7. Yoo, J.-Y., Ahn, S.-S., Seo, D.-H., Song, W.-H., Kang, K.-B.: New development of high grade X80 to X120 pipeline steels. Mater Manuf Process **26**, 154–160 (2011). https://doi.org/10.1080/10426910903202534
8. Cheng, S., Zhang, X., Zhang, J., Feng, Y., Ma, J., Gao, H.: Effect of coiling temperature on microstructure and properties of X100 pipeline steel. Mater Sci Eng A **666**, 156–164 (2016). https://doi.org/10.1016/j.msea.2016.04.066
9. Nafisi, S., Arafin, M.A., Collins, L., Szpunar, J.: Texture and mechanical properties of API X100 steel manufactured under various thermomechanical cycles. Mater Sci Eng A **531**, 2–11 (2012). https://doi.org/10.1016/j.msea.2011.09.072
10. Parisher, R.A., Rhea, R.A.: Pipe Fittings. Pipe Draft Des **13–55**, (2012). https://doi.org/10.1016/B978-0-12-384700-3.00003-7

11. Bursi, O.S., di Filippo, R., La, S.V., Pedot, M., Reza, M.S.: Probabilistic seismic analysis of an LNG subplant. J Loss Prev Process Ind **53**, (2018). https://doi.org/10.1016/j.jlp.2017.10.009
12. Corp. PD (2016) Mill Test Certificate No. 160818-FH01KS-0175A1–0001
13. (2013) ISO - ISO 17636–1:2013 - Non-destructive testing of welds — Radiographic testing — Part 1: X- and gamma-ray techniques with film
14. (2007) ISO - ISO 6520–1:2007 - Welding and allied processes — Classification of geometric imperfections in metallic materials — Part 1: Fusion welding

Detection of Cyber Attacks in Electric Vehicles Using a Deep Neural Network

Marcin Gajdzik[1,2,3](✉), Anna Timofiejczuk[1], Anna Gnacy-Gajdzik[1,2,3], and Piotr Przystałka[1]

[1] Department of Fundamentals of Machinery Design, Silesian University of Technology, 18a Konarskiego Street, 44-100 Gliwice, Poland
marcin.gajdzik@polsl.pl

[2] Doctoral School, Silesian University of Technology, 2a Akademicka Street, 44-100 Gliwice, Poland

[3] DIP Draexlmaier Engineering, Sp. z o. o., Gliwice, Poland

Abstract. The paper focuses on the identification of cyber-attacks in electric cars based on CAN bus signals anomalies detection. One of the objectives of the paper is to show that cyber-attacks may result in taking control over a module in an electric vehicle. On the other hand, the article suggests a protection method based on anomalies' detection using a deep neural network that can be used for increasing the level of CAN bus transmission security. This paper discusses an experiment that was conducted using an additional intruder module on CAN bus operating in bridge mode, which task is to change specific control signals in a way that is unnoticeable to the vehicle's supervisory system. The aim of the research is to analyse the time dependencies between the control signals transmitted on the CAN bus in a situation where a cyber attack on the vehicle has been carried out. By using deep neural learning algorithms, it is possible to detect anomalies suggesting the existence of a cyber-attack and to take countermeasures. It should be emphasised that the presented methodology for detecting anomalies of control signals on CAN buses can also serve as a tool of protection against taking control over the vehicle.

Keywords: cyber-attack · detection · deep neural network

1 Introduction

Since the first car left the assembly line, the automotive industry has experienced a remarkable transformation. More than 100 years of advances in technology have brought it to a place where electronic and IT solutions dominate this very mechanical discipline. Current top-end vehicles have about 200 million lines of code, up to 200 Electronic Control Units (ECUs) and more than 5 km copper wires [10]. In automotive engineering, more attention was paid to functionalities and safety than rather security before. In 2015, Charlie Miller et al. invaded the in-vehicle network of a Jeep by attacking the In-Vehicle Infotainment system (IVI), causing abnormal vehicle behavior [10]. Since

A. Puchalski et al. (Eds.): ICTD 2022, ACM 21, pp. 144–153, 2023.
https://doi.org/10.1007/978-3-031-31719-4_15

then, the public became interested in the automotive cybersecurity topic. The development of technologies like Information Communication Technology (ICT), Internet of Vehicles (IoVs), and industrial intelligence is gradually causing more intelligent and connected vehicles. Increased connectivity has provided us convenience and comfort, but simultaneously made vehicles more vulnerable to cyber-attacks. During the past 10 years, the frequent automotive cybersecurity incidents have revealed weaknesses in automotive cybersecurity [13]. The in-vehicle network, as an important component of Connected and Automated Vehicles (CAVs), is responsible for the crucial task of secure data transmission in the vehicle. It provides an external interface, resulting in vulnerability to potential security threats and attacks [3]. Security researchers have discovered that it is possible to exploit cyber-attacks by manipulating data on the internal vehicle control bus [4, 11]. The cyber-attack targeting this network may directly affect the functionalities of some safety-critical systems, such as the steering wheel, engine and brakes, which may endanger passengers or pedestrians [13]. In this context, the safety of modern vehicles is strictly related to addressing cybersecurity challenges [9]. To address these challenges, effective detectors and defense mechanisms need to be developed and evaluated for different types of attacks [12].

In this work, we proposed employing a deep neural network to detect potential cyber-attacks on the CAN bus. We conducted a research experiment by inserting an intruder module into the CAN network to modify the control signals on the bus imperceptibly to the vehicle. Several attacks were thus carried out during live driving of an electric vehicle around a test maneuvering area, logging all signals on the CAN bus simultaneously. The resulting log files, pre-processed using scripts written in Python, were used to train a deep neural network. The tests carried out confirmed the effectiveness of this network in detecting anomalies in the network unnoticeable to the vehicle, but which could endanger human health and life.

2 Background and Related Works

2.1 CAN Frame

The CAN (Controller Area Network) bus is an automotive standard used for reliable data exchange between electronic control units. At the physical layer a pair of twisted wires is used which is resistant to electromagnetic interference.

The main benefits of using CAN bus are: robustness, low-cost, speed, flexibility and efficiency.

The classical CAN was introduced in 1986. Over time, it has been extended to transmit increasing amounts of data. In the year 2015 the CAN FD became an international standard, known as ISO 11898–1. CAN FD is essentially a CAN bus with flexible data rate and an extension of the classical CAN data link layer. In comparison to classical CAN protocol, CAN FD increases the payload from 8 to 64 bytes. It also allows a higher data bit rate, depending on the CAN transceiver. CAN FD bus also requires additional terminating resistor with typical value 120 Ω at each end to eliminate reflection of electrical signals.

2.2 Attack Scenarios

There are two possibilities for adding the intruder module to the car CAN bus:

- parallel connected module to CAN bus - only receiving and sending data is possible, but no replacing frame content functionality is available.
- module as bridge is connected to the car CAN bus - receiving and sending frames is possible, but also blocking original ones and sending new with replace content in both directions.

2.3 A Bit Heat Map Concept

The idea of heat maps originated during reverse engineering work on finding control signals in an electric car and was presented by Marcin Gajdzik and Kamil Sternal [1].

To achieve this, a representation of the state of the bits in time must be drawn. An example of such a diagram is shown in Fig. 1. Each bit of the message is represented by a white pixel for a value of 1 or a black pixel when it is recognized as 0. In such a diagram, a single horizontal row of pixels represents the message content graphically represented in bits. Each subsequent pixel row represents the content of consecutive frames with the same identifier. This row with pixels is repeated sequentially for all captured frames in the vertical direction, from bottom to top. The resulting image in this way reveals which bits are active and helps to extract ranges of bits that could be potential signals involved in controlling the selected car feature.

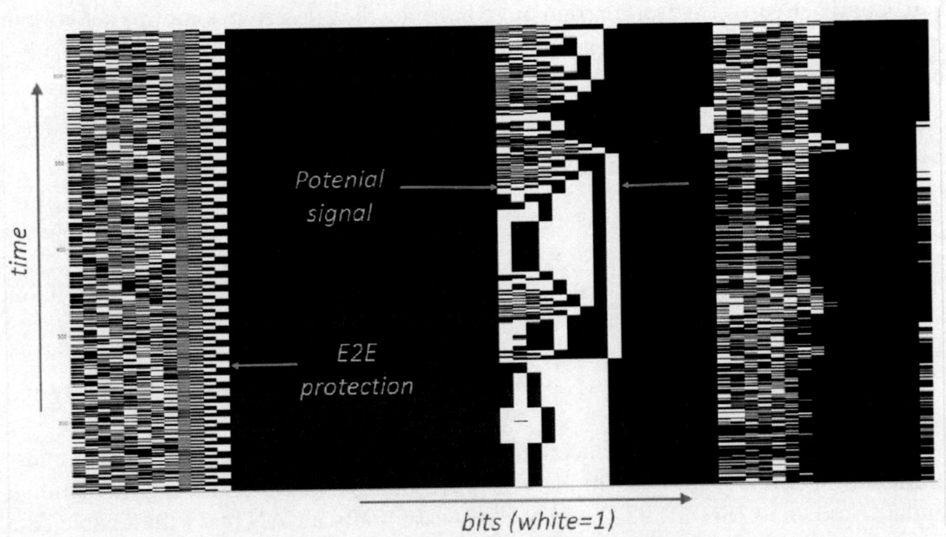

Fig. 1. Heat map with potential signals

3 Research Methodology

3.1 Intruder Module as a Bridge in Attack Experiment

Modern car models are complex mechatronic constructions. Dozens of electronic control units are connected to each other via bus systems, which are responsible for providing required functionality. The module that was chosen to attack and take control of the vehicle, was a power control steering unit, connected to the CAN bus. This was the focus of the experiment. From architecture perspective, there are two possible ways how to add intruder module to CAN bus.

First one is a parallel connection. In this configuration, it is not possible to replace the content of CAN frames. The only action that can be taken is to send an additional frame with the substituted content as soon as possible after receiving the original one (Fig. 1).

The second way is to connect the intruder module in series so that it acts like a bridge (Fig. 2). This gives you the possibility to block the original frame and send a new one with substituted content. During the research attack experiment, the second option was chosen.

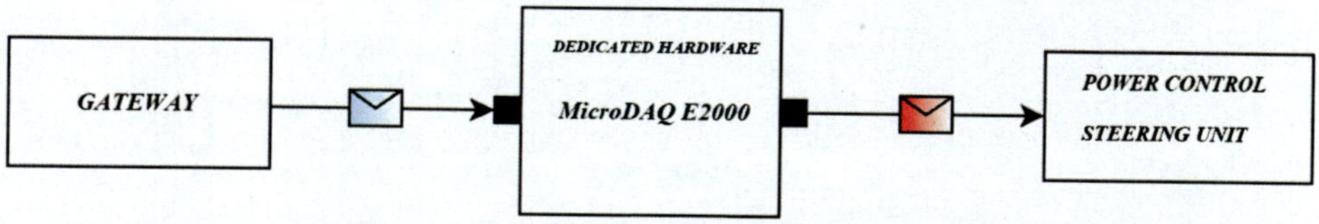

Fig. 2. Intruder module MicroDAQ E2000 as a bridge

The requirements for the test device working as an intruder module were very stringent. Especially in terms of the time needed to process a full CANFD frame protected by the End to End mechanism because the CRC would then have to be recalculated.

To meet this requirements, a dedicated MicroDAQ E2000 device was used for the experiment, performing the processing of a substituted CAN or CAN FD frame in less than one microsecond.

With this device, an attack was performed on the mangle by replacing the original content of the CAN frames with frames that had previously been logged during the operation of the lane assist system. This succeeded in taking control of the power steering unit.

3.2 A Bytes Map Concept for Anomaly Detection

Having collected several gigabytes of data logged from the CAN, to which power control steering module is connected, it was necessary to adapt it appropriately to the target format.

Based on the concept of a heat map, which represented successive bits in time for a given frame, thus enabling the detection of potential signals, a concept of bytes' map was proposed. The general idea is to show values of bytes with one millisecond resolution in time for selected CAN frames sequentially arranged (Fig. 3). The bytes from the data fields of the selected CAN frames were linked together according to the following order of identifiers: 0x40 (length = 8) + 0x86 (length = 8) + 0xFD (length = 8) + 0x116 (length = 8) + 0x176 (lcngth = 8) + 0x31B (length = 8) + 0x65D (length = 8) + 0x303 (length = 24) + 0xFC (length = 48) + 0x102 (length = 48).

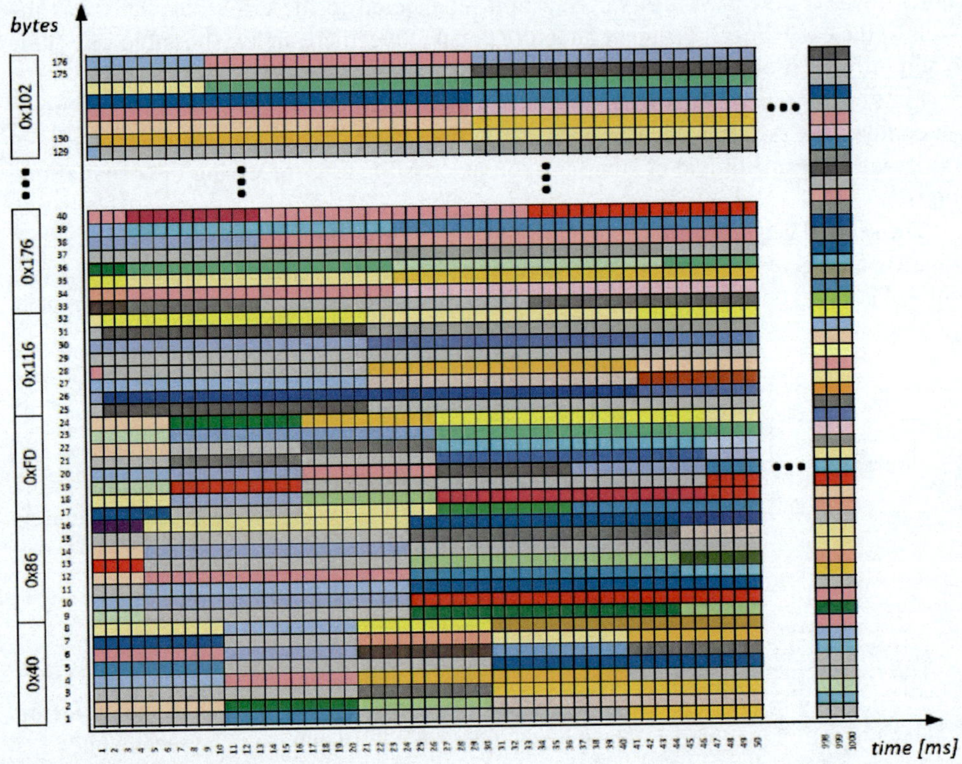

Fig. 3. A bytes map concept with assigned color bytes representation for one time window.

According to CAN standard, frames are sent on the bus cyclically with defined cycle time. Therefore, it is not possible to log CAN bus data for each frame each one millisecond. The value of cycle time is given in millisecond and strictly depends on the importance of the function performed by the signals contained in the frame. The more critical information is, the lower number of milliseconds is defined as a cycle time. Typical values of cycle time are between ten and one thousand milliseconds. In this case, to obtain a resolution in time equal to 1 ms, the collected data was processed using Python scripts. The data bytes for subsequent milliseconds were duplicated with the data received in the last received CAN frame. The proposed model represents the representation of bytes by colors - unique colors were assigned for values between 0 and 255. This representation in terms of bytes is comprehensible to humans, but is unacceptable when considering a sequence of bits understood as control signals. However, the model was prepared for deep neural networks, so it did not matter whether the data were represented as bits or bytes - they meant the same values fully understandable to the computer.

3.3 Anomaly Detection with Autoencoder Concept

An autoencoder is a type of deep learning network that is trained to replicate its input data. Autoencoders have surpassed traditional engineering techniques in accuracy and performance in many applications, including anomaly detection or image generation and recognition. By completing the encoding and decoding process, using a comparison of the original and reconstructed state, it is possible to deduce whether an anomaly has occurred that may be indicative of an ongoing attack (Fig. 4).

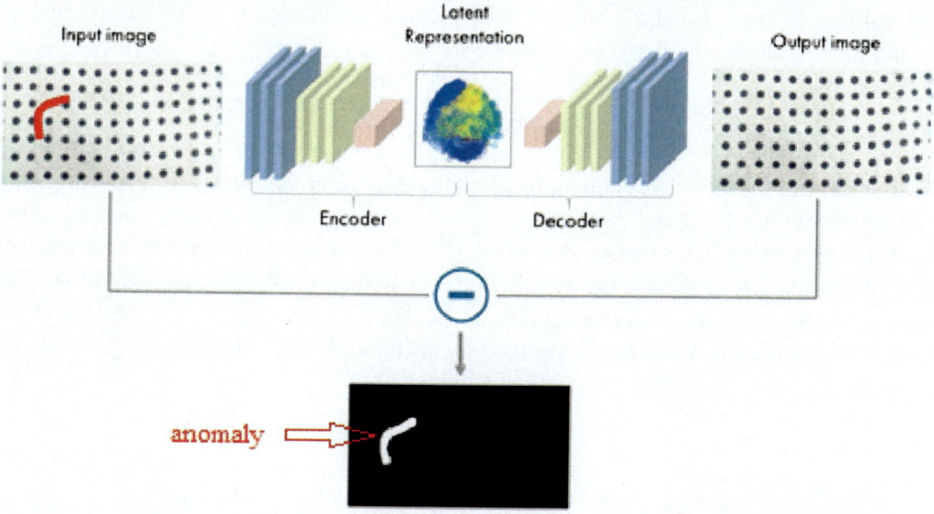

Fig. 4. Image bases anomaly detection with autoencoder [14].

Gated recurrent units auto-encoder (GRU-AE) neural networks, which were used in the experiment, are designed to consider the effect of previous state information on the current input data in the reconstruction of the original state. This makes it possible to realise the memorisation and reconstruction of anomaly data considering the temporal dependencies of the signals.

Taking advantage of the fact that the autoencoder is a type of network trained to reconstruct patterns that are given to it as input data, two types of input data were prepared with which the network was trained.

In the first case, the network was trained using data collected from several hours of driving the car in normal conditions, i.e. without attack and without using the lane assist system. So driver was the only responsible source for controlling power steering unit.

The second interesting case where the steering wheel control is not performed by the driver is the activation and use of the lane control system. Logged data from several hours of driving during which the lane assistant was active was used as an example to train the network when the vehicle takes over control of the system, but this is a fully intentional action.

The final stage of the experiment was to feed the data collected during takeover attacks to thc input of the trained deep neural network.

4 Results and Discussion

Having collected all the necessary data from the CAN bus in the defined normal cases, that is, when the steering wheel is operated only by the driver and when lane assist is used, they were entered into the proposed byte map model as valid states. With their use, a deep neural network of the GRU-AE type was trained. For this experiment bidirectional long-short term memory (BiLSTM) was used with 168 hidden units for bilstmLayer (total number of bytes) and 32 hidden units for gruLayer. However, this training was divided into 2 stages:

4.1 Deep Neural Network Training Without Lane Assist

The first stage of the study was conducted with the deep neural network trained using data collected during normal driving - that is, when the driver was the only one making turns with the steering wheel. As can be seen in the (Fig. 5), it was possible to accurately detect the anomaly that arises during the attack, that is, when control of the steering module is taken over. However, the difference between the signal indicating the anomaly and the signal at the output of the network for the correct case is small. The network should be retrained to get better results.

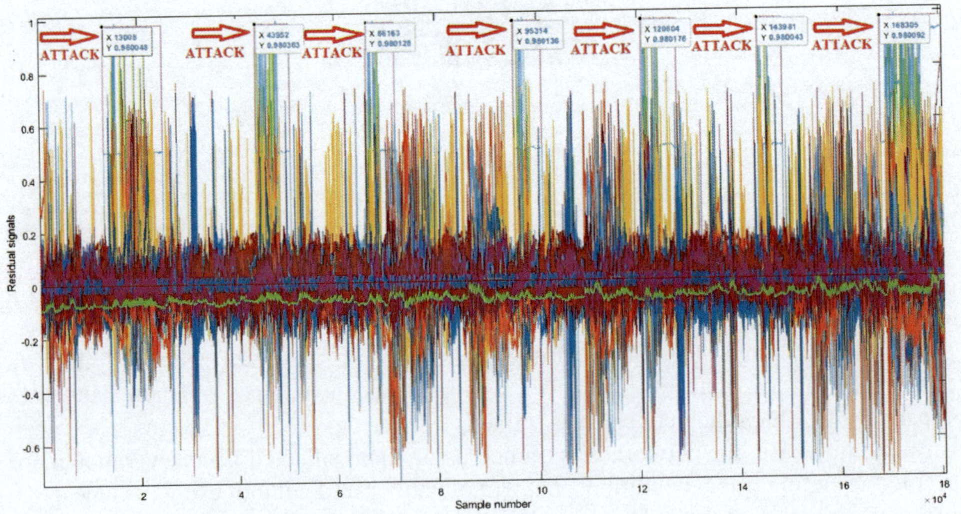

Fig. 5. Anomaly (attack) detection with DNN GRU-AE trained without lane assist

4.2 Deep Neural Network Training with Lane Assist

The second stage of the study was conducted with the deep neural network trained using data collected during normal driving - that is, when the driver was the only one making turns with the steering wheel and additionally when lane assist was active. As can be seen in (Fig. 6.), after training the network with data obtained while driving with active lane assist, anomaly detection became much easier. The difference between the signal at the time of the attack and the signal at the output of the network when given normal data is significant. This fact seems obvious, since the very same frames logged while driving with lane assist active were used for the attack.

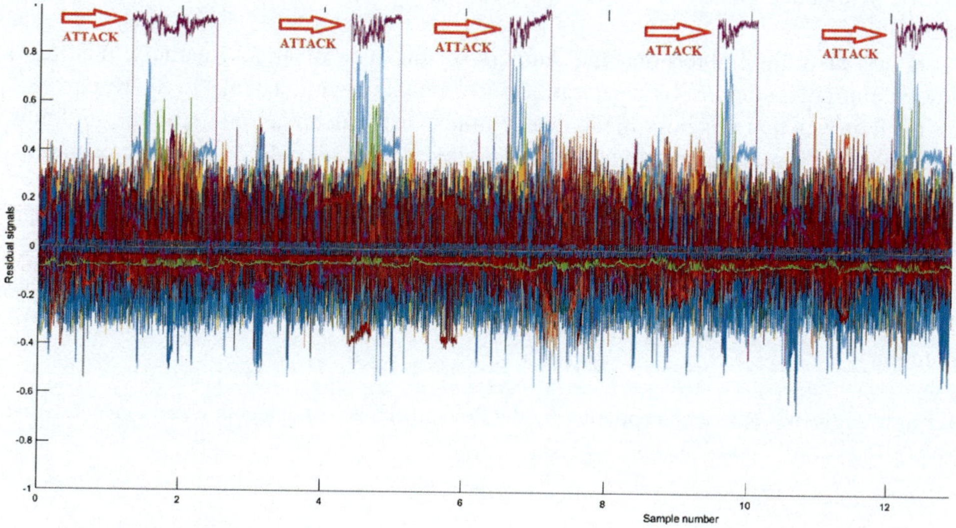

Fig. 6. Anomaly (attack) detection with DNN GRU-AE trained with lane assist

It is imperative to mention here that the possibility of taking control of the steering and turning suddenly with full torque can result in overturning of the vehicle and death of the passengers. Importantly, the intruder module can be unknowingly installed by the vehicle user himself, as there are many additional solutions on the market that allow remote control of the vehicle - just connect the purchased module to the bus.

5 Conclusion

As part of the research, several steps were carried out that were necessary to detect anomalies that could be information about an ongoing attack on the car. This included both logging data from the CAN bus, to which the power steering system is connected during normal driving, and driving using lane assist. And also collecting data during a successful attack related to taking control of the vehicle. In the experimental part of the research described in the paper, a real-world cyberattack scenario was conducted with the goal to take control of a function related to the safety of the car - in this case, the ability to turn. This helped to collect data from the bus during the attack.

The concept of a byte map, was proposed, i.e. consecutive CAN frames connected as data fields with a resolution of 1 ms and a time window of 1000 ms. The data processed with Python scripts was used to train the neural network.

Deep autoencoder neural networks using GRU were used as anomaly detection tool. Thanks to their properties of compressing the source data and then decompressing it along with taking into account time dependencies, it was possible to detect the attack carried out. Particularly noteworthy is the fact that the neural network trained only on data collected during normal driving was able to detect the anomaly when the vehicle was taken under control. However, the difference between the expected value at the output of the network and the anomaly was small. Nevertheless, if the network was additionally trained with data collected while driving with lane assist active, the difference was very noticeable.

It was also highlighted that the dangers of this type of attack, namely, making a sudden unpredictable turn by a car can end in a fatal accident. Thanks to the research, it has been proven that it is possible to detect attacks using deep neural networks, and thus prepare an appropriate scenario at the vehicle level, which would be able to prevent this type of situation or at least minimize its effects.

Further research development is planned using autoencoder neural networks to detect attacks on other electronic modules in the car, such as the braking system.

In addition, it is necessary to develop cases for teaching the network with portions of data from the same source but taking into account different time offsets on successive attempts.

Acknowledgment. Research supported by the Polish Ministry of Education and Science Grant No DWD/3/33/2019 and DWD/4/55/2020.

References

1. Cybersecurity, SAE J3061
2. Gajdzik, M., Sternal, K., Timofiejczuk, A., Przystałka, P.: A glimpse into the low-cost protection method dedicated for electric cars. In: 2021 5th International Conference on Control and Fault-Tolerant Systems (SysTol), pp. 72–78 (2021)
3. Koscher K., et al.: Experimental security analysis of a modern automobile, in 2010 IEEE Symposium on Security and Privacy, pp. 447–462: IEEE (2010)
4. Lee, H., Jeong, S.H., Kim, H.K.: Otids: A novel intrusion detection system for in-vehicle network by using remote frame In: Proceedings - 2017 15th Annual Conference on Privacy, Security and Trust, PST 2017, pp.57–66 (2017)
5. Luo, F., Zhang, X., Hou, S.: Research on Cybersecurity Testing for In-vehicle Network. Int. Conf. Intell. Technol. Embedded Syst. (ICITES) **2021**, 144–150 (2021)
6. Martínez-Cruz, A., Ramírez-Gutiérrez, K.A., Feregrino-Uribe, C., Morales-Reyes, A.: Security on in-vehicle communication protocols: Issues, challenges, and future research directions, Computer Communications, vol. 180, pp. 1–20, ISSN 0140–3664 (2021)
7. Oka, D.: Overview of the Current State of Cybersecurity in the Automotive Industry, in Building Secure Cars: Assuring the Automotive Software Development Lifecycle , Wiley, pp.1–21 (2021)
8. Road vehicles, Cybersecurity engineering, ISO/SAE 21434:2021

9. Scalas, M., Giacinto G.: Automotive Cybersecurity: Foundations for Next-Generation Vehicles. In: 2019 2nd International Conference on new Trends in Computing Sciences (ICTCS), pp. 1–6 (2019)
10. Simacsek, B: Can we trust our cars?, https://www.nxp.com/docs/en/white-paper/AUTOSECWP.pdf Last Accessed 26 Sept 2022
11. Song, H.M., Kim, H.R., Kim H.K.: Intrusion detection system based on the analysis of time intervals of can messages for in-vehicle network, Proceedings of the International Conference on Information Networking (ICOIN), IEEE, pp. 63–68 (2016)
12. Tariq, S., Lee, S., Kim, H.K., Woo, S., CAN-ADF: The controller area network attack detection framework, Computers & Security, Volume 94, 2020, 101857, ISSN 0167–4048
13. Upstream Security: 2020 Global Automotive Cyber security Report, Network Security, vol. 2020, Issue 1, 2020, p 4, ISSN 1353–4858 (2020)
14. https://www.mathworks.com/discovery/autoencoder.html, Last Accessed 26 Sept 2022
15. https://www.csselectronics.com/pages/can-fd-flexible-data-rate-intro, Last Accessed Sept 2022

The Diagnostic Procedure of Internal Characteristics Identification of the Semi-active Automotive Shock Absorber Using a Numerical Model

Szymon Żymełka(✉)

Silesian University of Technology, Gliwice, Poland
szymelka@polsl.pl

Abstract. In order to constantly improve automotive suspension, there are many new designs of shock absorbers that are developed every year. During such a development process, engineers have to measure and analyze multiple characteristics of the damper as well as its internal components. Many of those components, due to measurement instrumentation limitations (e.g. internal valves of the shock absorber) have to be checked at the isolated testing machines. Such an approach is expensive, limited to specific conditions and often requires the destruction of the sample shock absorber. The approach presented in the paper allows deriving those internal characteristics from the damper level tests which are much more accessible and easy to perform. The proposed method is based on the parameter identification procedure. To improve the performance of the optimization algorithm specifically for the problem of parameter identification a sensitivity-based approach is proposed. The performance of different sensitivity-based methods is compared on a benchmark case.

Keywords: Automotive · Shock absorbers · System modeling · Parameters identification Genetic algorithm · Sensitivity Study

1 Introduction and Problem Description

The shock absorber is a primary component influencing behavior of the suspension of modern cars. Due to its complexity, it is susceptible to various failure modes, many of which may compromise the safety of the passengers therefore before each new design of the shock absorber is released for production and application it has to be thoroughly tested and verified. During this process which usually takes around two years for a single design to complete, both external and internal characteristics of the damper are measured multiple times. External characteristics like damper force in the function of damper stroke and velocity can be relatively easily checked on the servo-hydraulic or electromagnetically

A. Puchalski et al. (Eds.): ICTD 2022, ACM 21, pp. 154–165, 2023.
https://doi.org/10.1007/978-3-031-31719-4_16

actuated, multipurpose testing machines (see Fig. 3). However internal characteristics have to be checked on dedicated testing equipment. [3] A notable example of such an internal characteristic that is measured frequently during shock absorber development the process is the flow rate vs pressure drop relationship of hydraulic valves (see Fig. 1).

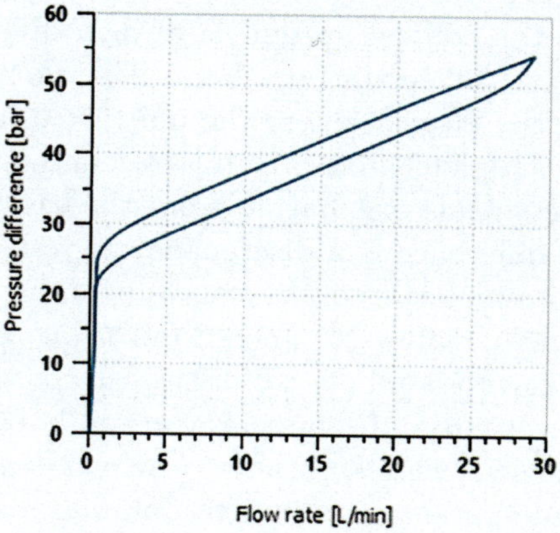

Fig. 1. Exemplary flow-rate vs pressure drop characteristic of a shim valve

This characteristic heavily influences the behavior of the shock absorber and is measured using component oriented test stand at which the flow is induced through the sample valve via two servo pumps and the pressure the differential is registered using a series of pressure gauges (see Fig. 2). Unfortunately, the measurement of this as well as other internal characteristics and parameters of the valve are only possible when the valve is extracted from the assembly of the shock absorber, meaning that from the perspective of the damper, this is a time-consuming and often destructive measurement method. Another disadvantage of the approach where the selected subsystems of the the damper is measured on the isolated test stand is that such an approach requires simplification of the operating conditions, for instance: the amount of gas dissolved and dispersed in the hydraulic medium might be different than in the real sample. Much more cost-effective and convenient for the development teams would be the possibility to establish approximated values of the internal characteristics directly from the external measurement of the assembled damper. As a result, there is a need for diagnostic parameters monitoring system capable of reproducing internally characteristics of the valve based on the force-displacement and force-velocity measurements of the damper. However, due to the fact that during the operation of the damper, multiple valves are employed at the same time, damper level forces can be reproduced for a given velocity with multiple combinations of the valves characteristics, therefore the task of establishing them is not possible using an analytical approach.

2 Proposed Solution

A proposed method of establishing the internal characteristics of the valves, based on the measurements of the damper assembly utilizes a numerical model parameter identification approach. At the beginning of this process, the same excitation signal should be supplied to the test and the simulation. In practice, the test is performed according to the target displacement signal which due to imperfection of the control system is always different from the actual displacement achieved by the testing machine. Therefore actual displacement signal measured during the test is sent to the numerical model. As a next step, after the simulation is complete, both results are compared by calculating the difference between force signals originating from the test and from the simulation. The result of this difference is a discrepancy signal, of which the absolute value is integrated to produce the total error value. The aim of the exercise is to minimize the total error value via an optimization algorithm by finding a correct set of input parameters and characteristics. In practice, the optimization surface is often very non-linear and some parameters that are subject to identification have a weak influence over the total error which is used as a goal function. On the other hand, some parameters tend to dominate the discrepancy signal. As a result of this situation, often a good agreement between the simulation and the experiment may be achieved even with not physically correct values of the parameters. In order to remedy the aforementioned difficulties, apart from the basic parameters identification approach explained above, sensitivity-based methods are proposed and their performance is compared using the benchmark case (see Sect. 4).

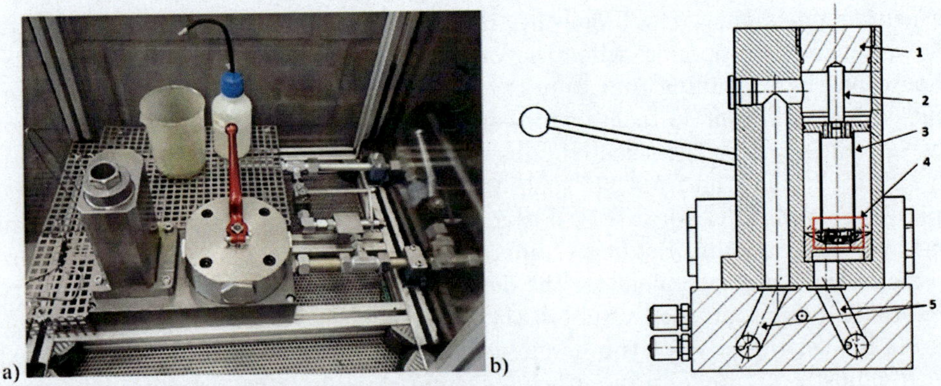

Fig. 2. . a) flow test bench, b) cross-section of the metal column: 1 - sealed nut, 2 - pin, 3 -inner tube, 4 - valve location, 5 - inflow/outflow channel

Since different parts of the shock absorber are active at different phases of the damper operation, the idea to improve the accuracy of the parameter identification is to involve in this process information about the sensitivity of the input parameters over the discrepancy signal. The first step of this approach requires performing sensitivity analysis [5,6] for the parameters that are being characterized. In the considered method, the first order Sobol index is to be calculated in function of time for each group of parameters related to a specific valve. The next step is to use the sensitivity signal to scale the discrepancy signal and therefore derive a better representation of the total error value for a given set of parameters. By having an error value more relevant for a specific set of parameters, it is possible to either use some of the multi-objective optimizations methods (in the benchmark case, the scale factors approach is used) or treat each sensitivity scaled error value as a separate goal function. In the former case which is the basis of the third investigated method, each group of parameters is identified by an independent genetic algorithm. However, despite sensitivity scaling, there is still a significant cross-talk between parameters of different groups due to second-order sensitivities. Because of that fact and also to reduce calculation time, genetic algorithms for each parameters groups are performed concurrently in such a way that each group of parameters creates only part of the chromosome of a given simulation run. When all the parts are available, the simulation is performed and the goal function is evaluated separately for each of the algorithms. Such an approach might be confusing to the underlying genetic algorithms since from the perspective of the singular algorithm, it loses its deterministic character but on the other hand, it might be possible to better utilize the sensitivity information (Fig. 4).

3 Model Description

Since the proposed method is heavily based on the numerical model it is imperative that mathematical formulas which are used, are capable of correctly reproducing physical phenomena that influence the outputs of interests. The following paragraph shortly describes the main layout of the numerical model. In the benchmark study, the model accepts the displacement signal as an input and outputs the signal of force. The value of force is determined based on the damper force balance (see Eq. 1). The pressure distribution used in the force balance equation is also an important part of the hydraulic system Eqs. 2 which are derived based on the law of the conservation of the mass. It involves changes in the volume of the chambers resulting from the imposed displacement, changes in the volume of the hydraulic medium due to compressibility and most importantly volumetric flow through the valves [1,2].

$$F_d = P_r A_r - P_c A_c + F_\mu \tag{1}$$

Fig. 3. Damper test set-up

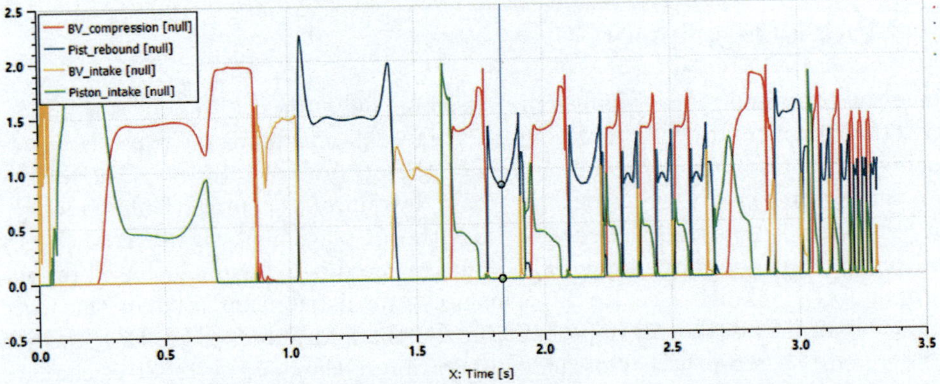

Fig. 4. Sensitivity functions for all parameters groups

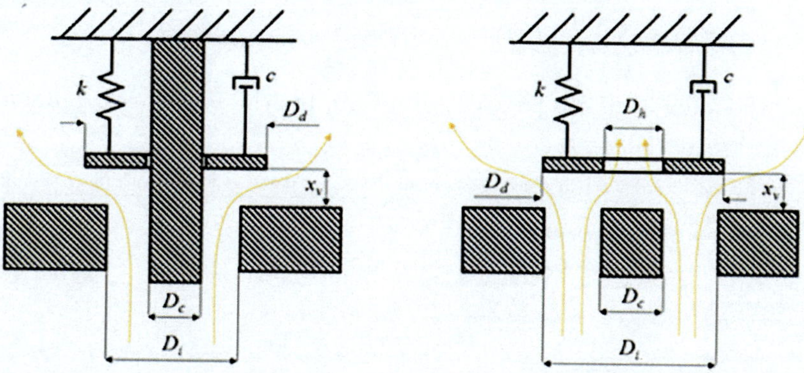

Fig. 5. Schematic drawing of the singular valve system

The relation between flow rate and pressure differential at the valve is determined for each valve using a separate mechanical system shown in Fig. 5. At each valve mechanical subsystem, the valve opening is calculated with the Eq. 3 which involves the following forces as valve disc force resulting from the pressure differential, frictional forces, and stiction force. When for a given pressure differential conditions the valve opening is established the flow through the valve is derived from the Eq. 4 [7,8].

$$\rho Q = \frac{\partial V}{\partial t} + \frac{V}{K}\frac{\partial p}{\partial t} \tag{2}$$

$$m_v(\ddot{x_v}) + c\dot{x_v} + F_k = F_{st} + F_{\Delta p} + F_{mu} + F_{mo} \tag{3}$$

$$Q = C_q A\sqrt{\frac{2\Delta P}{\rho}} \tag{4}$$

4 Benchmark Case

In order to verify the performance and accuracy of the proposed methods, benchmark parameters characterization process was performed for each method. Since the genetic algorithm used as an optimization algorithm (parameters listed in table 1) is inherently partially random [4], in order to have a measure of the consistency of the obtained characteristic, each algorithm was performed 5 times starting from different initial values. The following methods have been compared:

1. Baseline method - genetic algorithm using total error as a goal function
2. Sensitivity-based multi-objective approach where goal function has been defined according to eq. 5
3. Set of concurrent, independent genetic algorithms using sensitivity scaled goal functions for each passive valve

$$F(x) = \sum_{i=1}^{n} \int_{0}^{T} I_i(t) \cdot |F_{test}(t) - F_{simulation}(x_i, t)|\, \mathrm{d}t \tag{5}$$

where n is the number of the parameters group, I is the sensitivity signal and x_i is the i-th set of parameters.

Algorithm 1 Pseudocode of concurrently performed, sensitivity based genetic algorithms (method 3)

```
for each parameter_group g:
        Initialize population: Pg ← RandomIndividuals()
endfor
Combine chromosomes of all the individuals in population: P ← Combine(Pg1,Pg2,..,Pgn)
Run simulation for each individual to obtain result: R ← Simulations(P)
for each parameter_group g:
        Evaluate fitness: Fg ← Fitnessg (R) // fitness function uses its group's sensitivity signal
endfor
Initialize best solution: Ibest ← 0
while (given number of generations is reached)
        for each parameter_group g:
                Perform parents selection: Pg ← Selection(Pg,Fg)
                Perform crossover: Pg ← Crossover(Pg,Fg)
                Perform mutation: Pg ← Mutation(Pg,Fg)
        endfor
        Combine chromosomes of all the individuals in population: P ← Combine(P1g,P2g,..,Png)
        Run simulation for each individual to obtain result: R ← Simulations(P)
        for each parameter_group g:
                Evaluate fitness: Fg ← Fitnessg(R)
        endfor
        Save best solution in population: Ibest ← Fitness(R) //fitness function without sensitivity signal
endwhile
```

The parameters characterization process of the benchmark case was performed for the stiffness of each of the passive shock absorber valves (in total 4 valves). This parameter is primarily used to control the behavior of the shock absorber and is adjusted by selecting the correct configuration of the shim stack (see Fig. 6) or conical valve spring. Due to the contacts and non-uniform opening of the valve, this characteristic is strongly non-linear. The complete stiffness curve of each valve has been approximated by the function consisting of the three linear functions connected by a spline function (see Fig. 7). In total 8 parameters were used for the approximation of the valve stiffness.

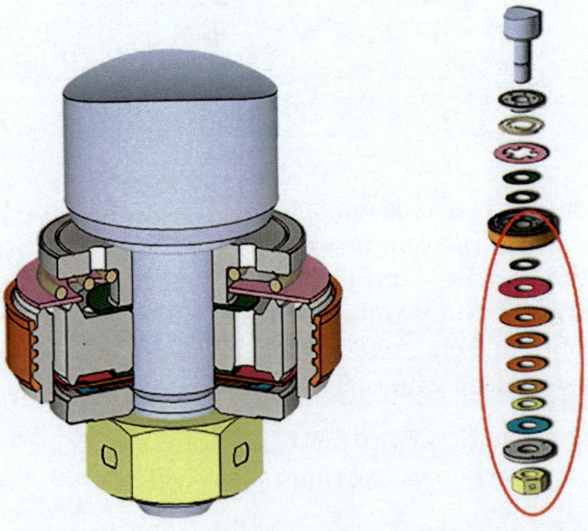

Fig. 6. View of the valve with the shim stack

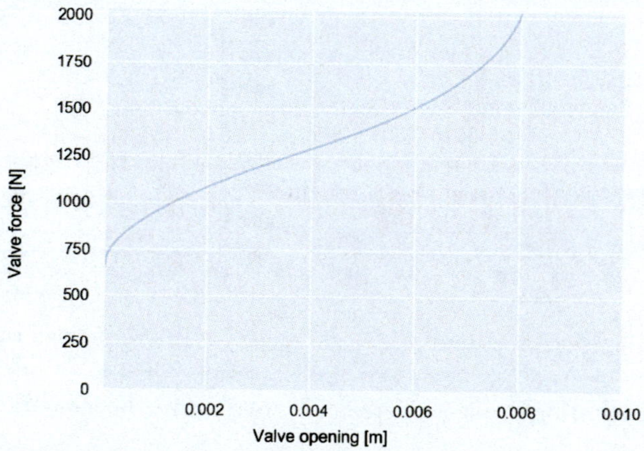

Fig. 7. Exemplary load-deflection characteristic of the valve

Table 1. Parameters of the genetic algorithm

Parameter name	Value
Number of generations	100
Number of parents	40
Solutions per generation	80
Cross-over type	Single point
Mutation type	Random
Mutation percentage	20
Parent selection type	Steady state

5 Results

The performance of the methods that were compared has been shown in Fig. 9a which illustrates the rate at which the algorithm can find an improvement to the goal function. As mentioned in Sect. 4, the algorithm is performed 5 times for each method. The best value of the goal function for each run can be observed in Fig. 9b and in Table 2. Two sample t-test that was performed for all the methods (see Table 3), indicates that the introduction of the additional sensitivity information used by the method, 3 brings a statistically relevant improvement in the best value of the objective function (for a given number of generations). The multiobjective approach which was based on the sensitivity signals failed to demonstrate an improvement in relation to the classical genetic algorithm.

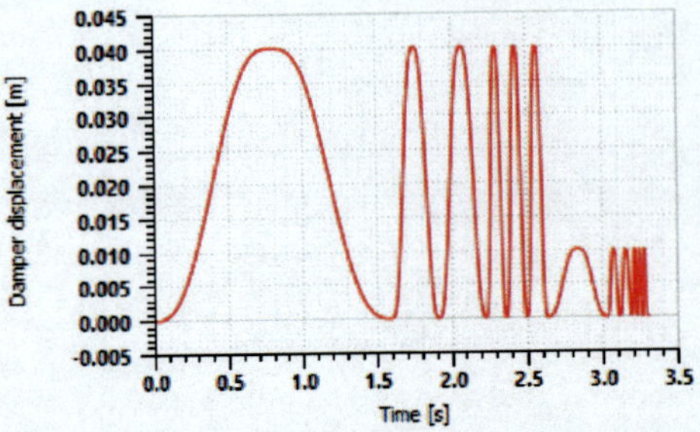

(a) Identification procedure

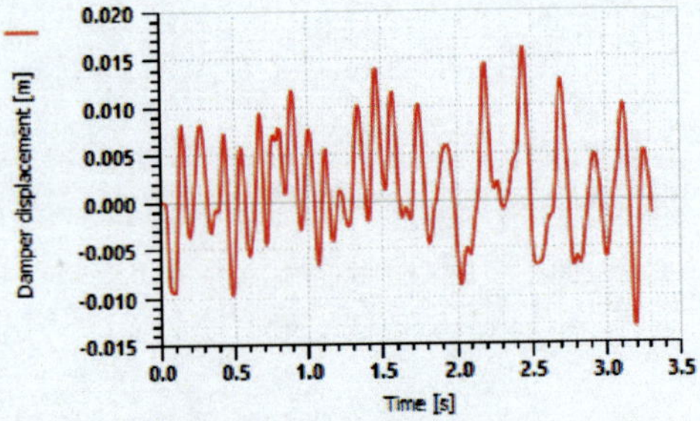

(b) Verifcation run

Fig. 8. Excitation displacement signals used during benchmark study

Table 2. Results of the identification run

Run no.:	Method 1	Method 2	Method 3
1	0.0722	0.0737	0.0787
2	0.0731	0.0703	0.0745
3	0.0693	0.0758	0.0792
4	0.0744	0.0747	0.0801
5	0.0734	0.0746	0.0743
Mean	0.0725	0.0738	0.0774
Std dev	0.0019	0.0021	0.0028

Table 3. Results of the t-test

Method 1	Method 2	t-value	p-value
1	2	0.0013	0.3276
1	3	0.0049	0.0119
2	3	0.0035	0.0532

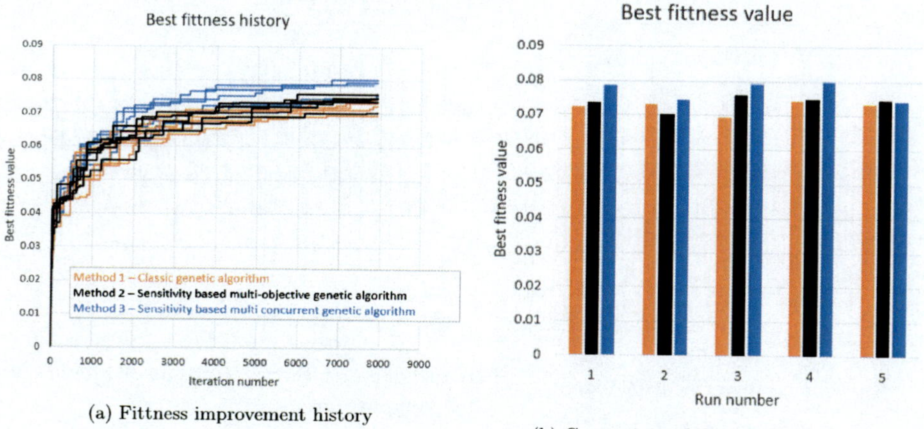

(a) Fittness improvement history

(b) Comparison of the best fitness values

Fig. 9. Comparison of the identification methods

Figures 10a and 10b show the comparison of the force from the experiment and simulations for method 3. In order to check whether parameters were not overfitted to the teaching data, an additional verification run has been performed. For this run, an unsprung mass noise signal was prescribed as an excitation (see Fig. 8b). A comparison of the forces from the verification run can be observed in Figs. 11a. To verify whether the model is capable to capture a higher frequency force outputs, the comparison is also performed for the high-pass filtered signal (Fig. 11b). Valve stiffness characteristics which established during the exercise (for each of the runs of the best performing method) have been shown in the Fig. 12.

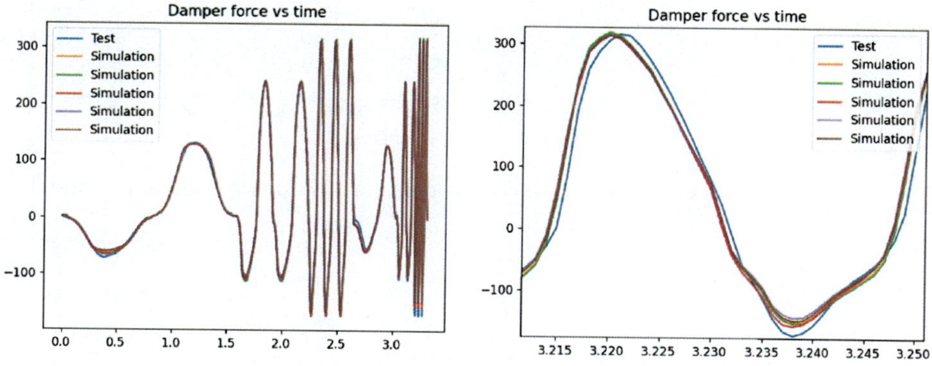

(a) Complete identification signal

(b) Close-up

Fig. 10. Comparison of the experimental and simulation force signal for the best iteration using method 3

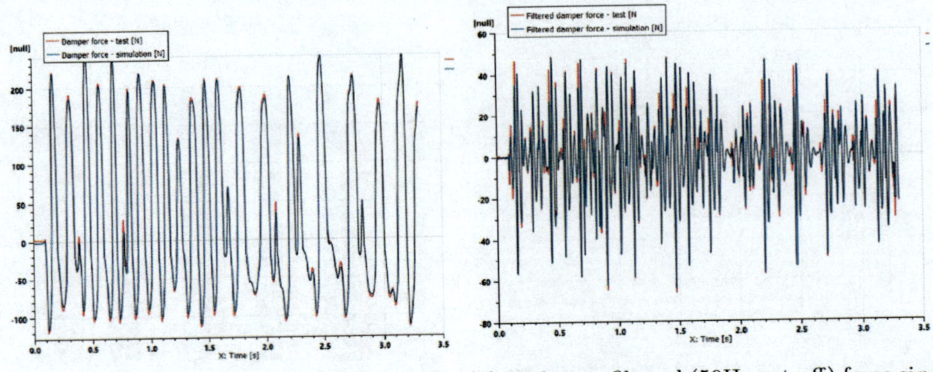

(a) Complete verification signal

(b) High pass filtered (50Hz cut-off) force signal comparison

Fig. 11. Comparison of the experimental and simulation force signal for the verification signal

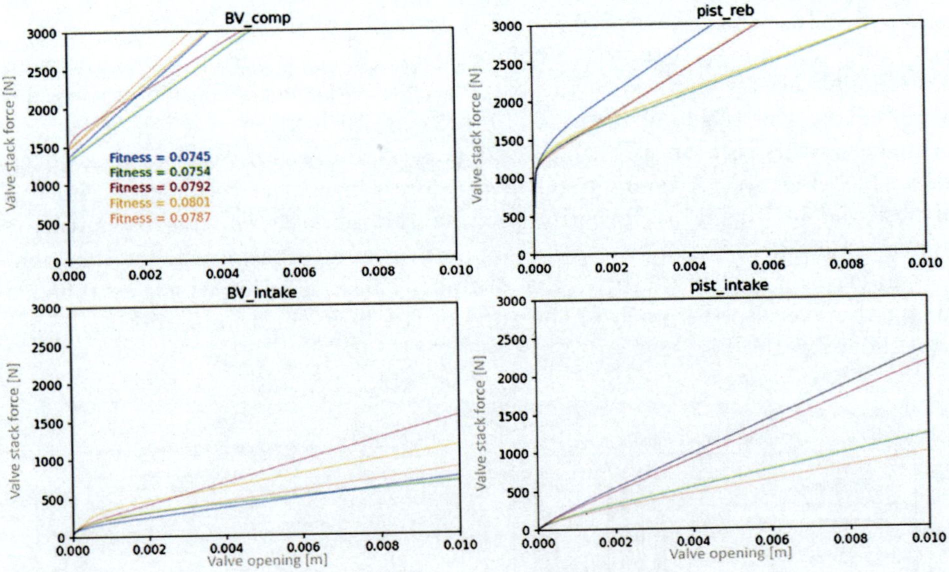

Fig. 12. Valve stiffness characteristics after identification process using method 3

6 Conclusions

– Inspection of the Fig. 11a allows concluding that the parameters of the model have been characterized well enough to closely reproduce the behavior of the shock absorber. This conclusion is additionally proved by the validation run which confirms that the parameters were not overfitted to the supplied teaching dataset. Comparison of high-pass filtered force signal (Fig. 11b) shows that the model results using identified parameters are also valid at higher frequencies.

- Stiffness of the valves which constitutes a key internal characteristic of the damper has been approximately identified although the scatter remains relatively high for the considered settings of the optimization algorithm. In order to improve the accuracy, the number of generations of the genetic algorithm could be increased (in the benchmark case, it was relatively low since the algorithm was repeated 5 times). This creates also a need for a procedure for determining when the desired level of accuracy is reached for a given identification procedure.
- Sensitivity-based method allowed for a noticeable improvement in the performance of the optimization algorithm. Concurrently run set of genetic algorithms with sensitivity scaled goal functions performed the best out of considered methods. Further improvement could be achieved by adjusting genetic algorithm parameters as well as methods of defining sensitivity signals.

Acknowledgements . The research was co-financed under grant no. RJO15/SDW/001-61/US supported by the Ministry of Science and Higher Education in Poland. Additionally, the author is grateful to the Tenneco Automotive Eastern Europe Sp. z o.o. for the good cooperation and the opportunity to publish the research results.

References

1. Benaziz, M., Nacivet, S., Thouverez, F.: A shock absorber model for structure-borne noise analyses. J. Sound Vibr. **349**, 177–194 (2015)
2. Czop, P., SLawik, D.: A high-frequency first-principle model of a shock absorber and servo-hydraulic tester. Mech. Syst. Signal Process. **25**(6), 1937–1955 (2011)
3. Dixon, J.C.: The Shock Absorber Handbook (1999)
4. Gad, A.F.: PyGAD: an intuitive genetic algorithm python library, p. 7 (2021)
5. Herman, J., Usher, W.: SALib: an open-source python library for sensitivity analysis. J. Open Source Softw. **2**(9), 97 (2017)
6. Iwanaga, T., Usher, W., Herman, J.: Toward SALib 2.0: advancing the accessibility and interpretability of global sensitivity analyses. Socio-Environ. Syst. Model. **4**, 18155 (2022)
7. Lang, H.H.: A Study of the Characteristics of Automotive Hydraulic Dampers at High Stroking Frequencies. PhD thesis (1977)
8. Sikora, M.: Selected Aspects of Dynamic Behaviour of Automotive Hydraulic Shock Absorbers. PhD thesis (2021)

Check for updates

Fault Diagnosis Using Artificial Neural Networks Trained Only on Signals from an Undamaged Machine

Paweł Pawlik[1](✉), Konrad Kania[2], and Bartosz Przysucha[2]

[1] Department of Mechanics and Vibroacoustics, AGH University of Science and Technology, al. A. Mickiewicza 30, 30-059 Kraków, Poland
pawlik@agh.edu.pl

[2] Department of Quantitative Methods in Management, Lublin University of Technology, Ul. Nadbystrzycka 38 D, 20-618 Lublin, Poland

Abstract. Fault diagnosis of machines operating under variable conditions requires advanced signal analysis methods. Variable conditions of load, temperature or rotational speed influence the values of vibration signals.

This paper proposes a diagnosis method based on order analysis and an artificial neural network trained solely on data for a machine operating in a fault-free condition. The order spectrum of the new parameter rDPNS, which does not depend on the machine's working conditions, was proposed. The obtained order spectrum of this parameter allows the identification of faults by the theory of fault diagnostics.

The proposed method has been verified in diagnosing the degradation of a two-stage cylindrical gear. The diagnostic experiment was conducted on a laboratory bench. The signals of vibration acceleration, rotational speed, and current supply to the drive motor were recorded at varying load and temperature.

The results of the experiment conducted on the laboratory bench showed the effectiveness of the proposed method.

Keywords: fault diagnosis · vibroacoustic diagnostics · neural networks · deep learning · degradation process

1 Introduction

In industry, there is a demand for diagnosing machines operating under variable conditions. Variable operating conditions include varying load, temperature and rotational speed.

Changing the load affects the speed of the machine and the values of measured vibrations [1–3]. Synchronous methods are used, which allow the vibration signals to be synchronized with the rotation of the machine [4– 8]. However, these methods do not solve the problem of influencing the amplitude of vibrations.

In paper [1], a method describing the relationship between order amplitudes and load in the form of polynomial curves is presented. Differences between the polynomial

A. Puchalski et al. (Eds.): ICTD 2022, ACM 21, pp. 166–175, 2023.
https://doi.org/10.1007/978-3-031-31719-4_17

curves are presented as diagnostic parameters. The paper [9] proposes a novel approach to analysing cyclo-non-stationary signals based on generalising cyclo-non-stationarity indices to cover simultaneous and independently varying speed and load operating conditions. However, the values of the vibration amplitude are also influenced by the oil temperature as shown in work [10]. Analysis of the order spectrum becomes difficult if a variable load and temperature simultaneously exists. In the paper [11], the authors attempted to solve this task by using deep learning methods to describe these relationships. However, the set of neural networks used there is extensive and complicated. The problem of choosing the right architecture has been studied for years [12, 13]. To this day some new papers on the subject are still appearing, considering new types of architectures in the context of specialized applications [14].

This thesis presents the use of an artificial neural network with a relatively simple architecture to diagnose machines operating under varying conditions. The proposed method to train the neural network requires only data from an undamaged machine and the proposed novel order spectrum allows fault identification.

2 Methodology

2.1 General Description of the Novel Method

The method is used to diagnose damage to machinery operating under varying conditions. Vibration, rotational speed and operating conditions signals are inputs to the method. The signals are then subjected to a pre-processing algorithm. The vibration signals are synchronized with the machine's rotational shaft speed signal. In the next step, the orders' amplitudes as a load function are determined, which, together with signals describing the machine's operating conditions, are fed to the input of the neural network-based algorithm.

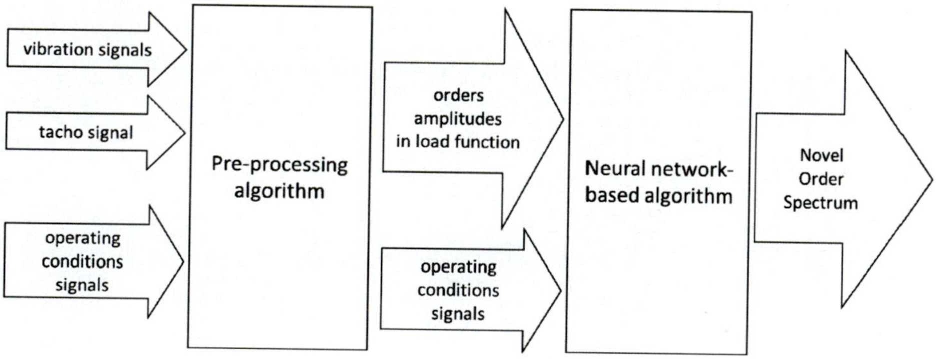

Fig. 1. Novel method algorithm

Thus an important aspect of presented method is that the artificial neural network is trained only with data from an undamaged machine. Then, in the diagnosis process, the actual measured data is fed to the algorithm's input based on the trained neural network.

The output of the method is a novel order spectrum that does not depend on machine conditions. The algorithm of the method is illustrated in Fig. 1.

2.2 Pre-processing Algorithm

The method involves measuring vibrations and rotational speed, current and temperature simultaneously. The rotational speed signal is used in the Order Analysis method to synchronize vibration signals with varying machine rotations. On the other hand, the current signal carries information about the load.

First, the waveforms of order amplitudes, current and temperature as a load function are determined. Then a moving averaging operation is carried out. The data thus obtained is the input to the neural network algorithm.

2.3 Neural Network-Base Algorithm

In order to overcome the problem of variational operating conditions and the issue of distinguishing whether changes in orders amplitudes come from a change in conditions or the appearance of a fault the specific neural network-based algorithm was develop.

The algorithm consists of two main parts:

1. An actual model of regression neural network (RNNM) that maps relations between amplitudes of orders and operating conditions signals in a case of undamaged machine.
2. The procedure of creating a novel order spectrum based on the network response error analysis in the context of the impact of each order to that error.

The concept of the whole procedure is presented in Fig. 2.

The key aspects of used neural network is that it was trained only on data from undamaged machine to estimate a mapping from orders amplitude to operating conditions provided that the machine is healthy.

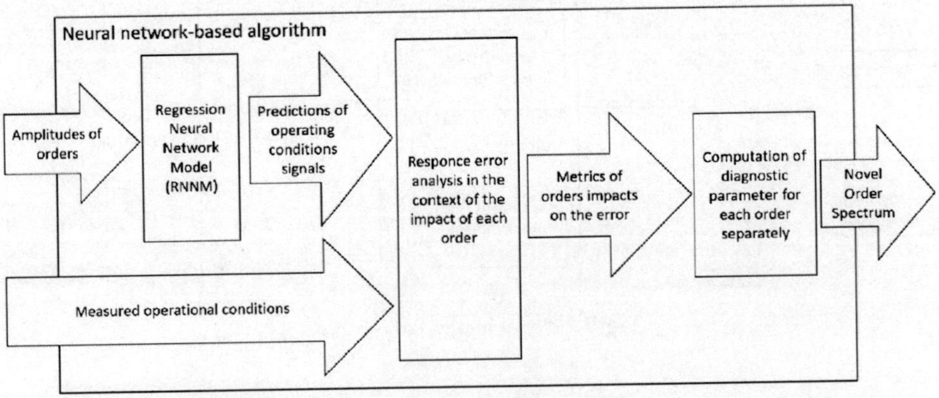

Fig. 2. Scheme of neural network-based diagnostic algorithm

The actual architecture was built upon deep neural network based on fully-connected layers.

The actual diagnostic process was based on the assumption that a fault in a machinery causes a change of the relationships between signals generated by the machine. Due to the fact that neural network was trained only on signals from the undamaged machine, it can be used as a tool to observe and quantify the levels of that changes in signals relations what can be utilized directly in a diagnostic process. If relations of machinery signals remain the same, we can expect that the range of the response error of the RNN model also remain unchanged. However, as the relations between machinery signals are disturbed through the beginning of formation of a fault the disturbance of the response error of RNN should be noticeable as well.

2.4 Novel rDPNS Spectrum

While the mere observation of the error makes it possible to detect the fault, our goal was to provide a method also for identifying the type of the fault which could be used for automated diagnostics and at the same time would return information understandable to an experienced diagnostician. In order to achieve that aim we determine the impact on response error of RNN for each order in spectrum separately.

To measure that impact for specific order we use the following procedure:

1. Compute the responses of RNN for train data (from undamaged machine)
2. Swap the amplitudes of chosen order in training data with the corresponding amplitudes from healthy machine but taken from test data. (unused in training process)
3. Compute the Euclidean norms of differences of responses before and after the swapping.
4. Compute the mean (m_g) and the variance (${s_g}^2$) of computed norms.
5. Repeat steps 1–4 but instead of training data use the data that are checked for faults to get yet another mean (m_f) and variance (${s_g}^2$).

We can now introduce the novel parameter Relative Differences Product of Network Statistics (rDPNS). For case where the swapping was performed on the ith order the rDPNS for that particular order is given by the formula:

$$rDPNS(i) = ReLU\left(\frac{s_f^2(i) - 2s_g^2(i)}{s_f^2(i) + 2s_g^2(i)}\right) \cdot ReLU\left(\frac{m_f(i) - m_g(i)}{m_f(i) + m_g(i)}\right) \tag{1}$$

$$ReLU(x) = \begin{cases} x \text{ } if \text{ } x \geq 0 \\ 0 \text{ } otherwise \end{cases} \tag{2}$$

Notice that the role of the ReLU function here is just to set rDPNS parameter to zero if any (or both) of the differences in the formula are negative which we interpreting as no evidence for the appearance of the fault in examinated data.

Moreover, as differences in formula for rDPNS are relative the whole parameter is bounded to the interval [0,1]. Due to that boundness of rDPNS to classify the strength of the observed effect we can utilize the Cohen scale [15] (Table 1).

Table. 1. Cohen Scale.

Range of parameter	The strength of the effect
0–0.1	No effect
0.1–0.3	Little effect
0.3–0.5	Medium effect
0.5–1.0	Large effect

The rDPNS measure is a measure normalized to 1 that simultaneously tests the dispersion of average values and compares the increase in variance in the signals being compared. Since the value of this parameter is normalized, a standard scale can be applied to the measure of effect strength. In this case, the Cohen scale was chosen, which is originally used in the social sciences, but can also be applied to parameters in the technical sciences.

As we compute the rDPNS for all available orders and classify the rDPNS values according to the Cohen scale we can established (even automatically) if there is some fault or not. Presented method of obtaining the spectrum based on rDPNS guarantees insensitivity to the variational operation conditions. This feature allows the identification of the type of damage detected according to the classical theory of order spectrum analysis.

3 Experimental Validation

3.1 Test Rig

To verify the proposed method, an experiment was conducted on a laboratory rig (see Fig. 3). The bench allows you to task any gearbox load function. The test object was a two-stage helical gearbox. The characteristic frequencies and orders are shown in (Table 2).

Table 2. Characteristic orders and frequencies of the tested gearbox.

	Order [-]	Frequency [Hz]
Output shaft	1	11.8
Input shaft	3.7	43.66
First stage gear meshing	56	660.8
Second stage gear meshing	155.55	1835.49

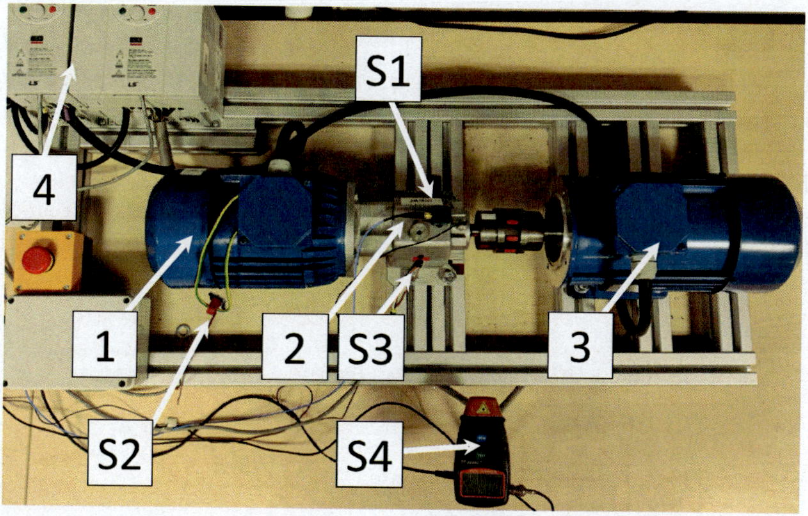

Fig. 3. Test rig, 1) drive motor, 2) gearbox, 3) load motor, 4) converters S1) accelerometer, S2) current sensor, S3) temperature sensor, S4) tachometer.

3.2 Experiment

The experiment consisted of removing the lubricating oil from the gearbox and then exploring for eight days. During the investigation, the acceleration of vibrations on the gearbox body was recorded using PCB 356B08 accelerometer. The rotational speed, temperature and current consumed by the driving motor were measured simultaneously.

The gearbox was loaded with a variable torque corresponding to the load on the bucket excavator's main gearbox. A constant frequency voltage powered the drive motor. So any speed changes were caused by a change in load.

Figure 4 shows the interior of the gearbox before and after degradation. Discolouration on the gears caused by overheating can be seen. The profile of the teeth has also changed due to blurring.

Fig. 4. Gearbox interior. Before degradation (left), after the 8th day of degradation (right).

3.3 Order Analysis Results

On the 4th day of degradation, a slight increase in the amplitudes of the mesh components is visible (see Fig. 5). An interesting phenomenon is that the signal for the damaged machine is at smaller values than the signal for the undamaged machine. This may be due to the initial arrival of the new machine.

Due to the varying load and temperature, a significant variance of order amplitudes is visible. For machines operating under variable loads, concluding damage is difficult because the values strongly depend on the load.

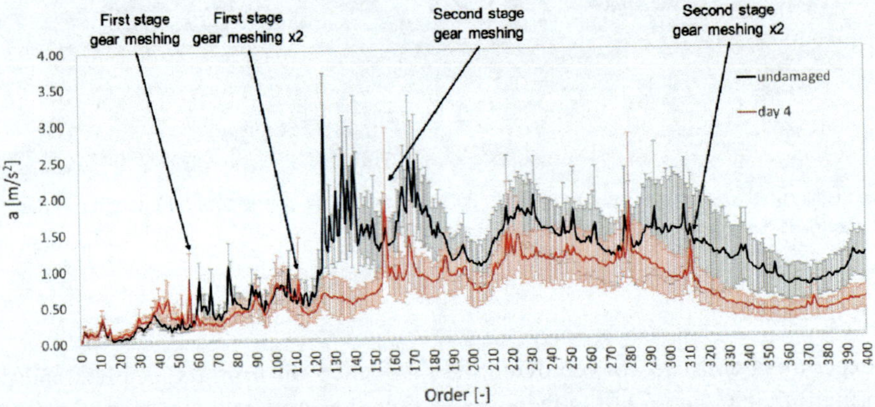

Fig. 5. Order spectrum of vibration signal, undamaged machine (black), machines after the 4th day of operation (red).

On Day 8, the values of mesh orders are higher (see Fig. 6). However, these values are not significant compared to the spectrum for an undamaged machine (Fig. 6).

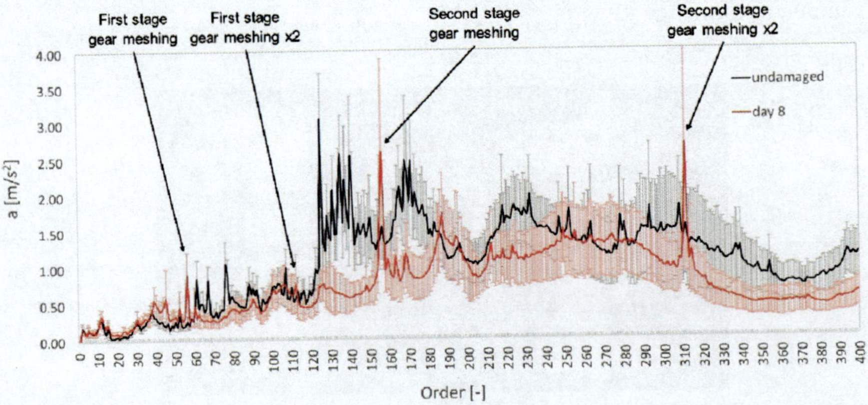

Fig. 6. Order spectrum of vibration signal, undamaged machine (black), machines after the 8th day of operation (red).

In conclusion, analysis of the averaged order spectrum is difficult due to the deviations of amplitudes caused by varying load and temperature.

3.4 Neural Network-Based Algorithm Results

Here we present the results of described earlier novel spectrum of rDPNS parameter. (Fig. 7) shows the spectrum for the 4th day of the experiment, while Fig. 8 is the spectrum for the 8th day of operation without oil. Horizontal lines on plots are the thresholds of Cohen scale (yellow: 0.3 level, red: 0.5 level).

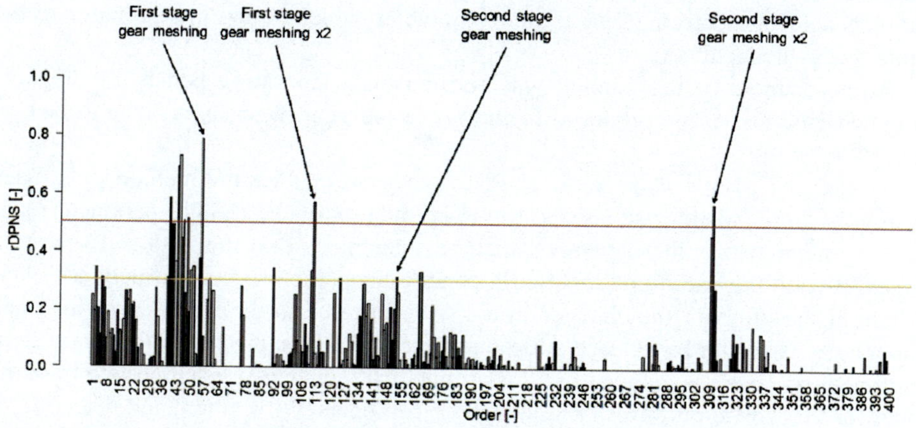

Fig. 7. rDPNS Spectrum after the 4th day of operation.

The Cohen scale makes interpretation of the rDPNS spectrum more accessible and more usable for automated diagnostic systems. On the 4th day, values for the first and second stage mesh order above the 0.5 level (large effect) can be observed (see Fig. 7).

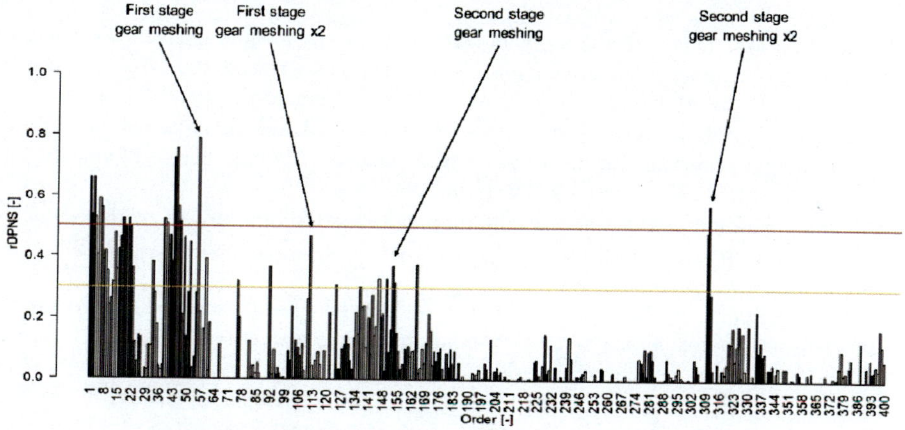

Fig. 8. rDPNS Spectrum after the 8th day of operation.

For day 8, the situation is similar. Still, even lower components appear, corresponding to the gearbox's input and output shaft rotations (see Fig. 8).

In summary, the proposed novel spectrum is more accessible to interpret than the classic order spectrum, which is sensitive to changes in load and temperature.

4 Summary

The paper presents a novel method for diagnosing rotating machinery operating under varying conditions. A novelty in the proposed method is the way to utilized an artificial neural network trained only with data from the undamaged machine. The output of the method is a novel spectrum of the rDPNS parameter, which allows interpretation of the results by diagnostic theory.

The correctness of the method was verified on a laboratory bench for diagnosing cylindrical gears. The proposed method detected gear mesh failure due to lack of lubrication.

The use of Cohen's scale in the rDPNS Spectrum enables the method to be used in automated diagnostic systems due to the boundness of the rDPNS parameter. It is worth to mention that the developed method guarantees that the obtained spectrum is insensitive to the variational operation conditions as the used neural network plays the role of the filter for the changes in the spectrum due to the variation of load and temperature. It should be noted that despite supervised learning in the form of a deep neural network, there is no need for a priori data from damaged machinery to train the model for diagnostic purposes. The technique can be used for automated detection of any kind of fault and provide information about the damage in the form of the spectrum.

Acknowledgements. This work was supported by the Polish Ministry of Science and Higher Education [grant number 16.16.130.942].

References

1. Pawlik, P.: Single-number statistical parameters in the assessment of the technical condition of machines operating under variable load. Eksploatacja i Niezawodnosc – Maintenance and Reliability. **21**, 164–169 (2019). https://doi.org/10.17531/ein.2019.1.19
2. Pawlik, P., et al.: Vibroacoustic study of powertrains operated in changing conditions by means of order tracking analysis. Eksploatacja i Niezawodnosc – Maintenance and Reliability. **18**, 606–612 (2016). https://doi.org/10.17531/ein.2016.4.16
3. Bartelmus, W., Zimroz, R.: A new feature for monitoring the condition of gearboxes in non-stationary operating conditions. Mech Syst Signal Process. **23**, 1528–1534 (2009). https://doi.org/10.1016/j.ymssp.2009.01.014
4. Burdzik, R., Konieczny, Ł, Warczek, J., Cioch, W.: Adapted linear decimation procedures for TFR analysis of non-stationary vibration signals of vehicle suspensions. Mech Res Commun. **82**, 29–35 (2017). https://doi.org/10.1016/j.mechrescom.2016.11.002
5. Cheng, W., Gao, R.X., Wang, J., Wang, T., Wen, W., Li, J.: Envelope deformation in computed order tracking and error in order analysis. Mech Syst Signal Process. **48**, 92–102 (2014). https://doi.org/10.1016/j.ymssp.2014.03.004

6. Wu, K., Xing, Y., Chu, N., Wu, P., Cao, L., Wu, D.: A carrier wave extraction method for cavitation characterization based on time synchronous average and time-frequency analysis. J Sound Vib. **489**, 115682 (2020). https://doi.org/10.1016/j.jsv.2020.115682
7. Wang, W., Johnson, K., Galati, T.: Vibration analysis of planet gear bore-rim failure using enhanced planet time synchronous averaging. Eng Fail Anal. **117**, 104942 (2020). https://doi.org/10.1016/j.engfailanal.2020.104942
8. Zhang, S., Tang, J.: Integrating angle-frequency domain synchronous averaging technique with feature extraction for gear fault diagnosis. Mech Syst Signal Process. **99**, 711–729 (2018). https://doi.org/10.1016/j.ymssp.2017.07.001
9. Mauricio, A., Helm, D., Timusk, M., Antoni, J., Gryllias, K.: Novel Cyclo-Nonstationary Indicators for Monitoring of Rotating Machinery Operating Under Speed and Load Varying Conditions. J Eng Gas Turbine Power. **144**, (2022). https://doi.org/10.1115/1.4049778/1096343
10. Pawlik, P.: The diagnostic method of rolling bearing in planetary gearbox operating at variable load. Diagnostyka. **20**, 69–77 (2019). https://doi.org/10.29354/diag/111567
11. Pawlik, P., Kania, K., Przysucha, B.: The Use of Deep Learning Methods in Diagnosing Rotating Machines Operating in Variable Conditions. Energies (Basel). **14**, 4231 (2021). https://doi.org/10.3390/EN14144231
12. Wilamowski, B.M.: Neural network architectures and learning algorithms. IEEE Ind. Electron. Mag. **3**, 56–63 (2009). https://doi.org/10.1109/MIE.2009.934790
13. Benardos, P.G., Vosniakos, G.C.: Optimizing feedforward artificial neural network architecture. Eng Appl Artif Intell. **20**, 365–382 (2007). https://doi.org/10.1016/J.ENGAPPAI.2006.06.005
14. Ilina, O., Ziyadinov, V., Klenov, N., Tereshonok, M.: A Survey on Symmetrical Neural Network Architectures and Applications. Symmetry **14**, 1391 (2022). https://doi.org/10.3390/SYM14071391
15. Cohen, J.: Statistical Power Analysis for the Behavioral Sciences, 2nd edn. Lawrence Erlbaum Associates, New York (1988)

Algorithm of Sensory Network Routing Operating in Explosion Hazard Zones

Jerzy Jagoda[1(✉)], Krzysztof Stankiewicz[1], Joanna Rogala-Rojek[1], and Mariusz Hetmanczyk[2]

[1] Department of Mechatronic Systems, ITG Komag, Pszczyńska 37 St., 44-100 Gliwice, Poland
jjagoda@komag.eu

[2] Faculty of Mechanical Engineering, Department of Process Automation and Integrated Manufacturing Systems, The Silesian University of Technology, Konarskiego 18A St., 44-100 Gliwice, Poland

Abstract. This paper presents the author's concept of a routing algorithm for a sensor network operating in methane and/or coal dust explosion hazard zones. In such extensive networks, it is necessary to organise the transmission of information appropriately in order to ensure operational reliability, flexible configuration and operational safety. It is particularly important to ensure the operational reliability of sensor networks operating in methane and/or coal dust hazardous areas. The result of the analysis of individual routing protocols, indicate that protocols based on the so-called Swarm Intelligence (SI) have the highest probability of successful algorithm implementation in sensor networks operating in explosive environments. These algorithms are characterised by the fact that, regardless of the number of variables in the solution space, they adapt to the constraints and help solve the optimisation problem. Sensory networks working on the basis of swarm intelligence algorithms are self-organising and multiredundant. The paper also presents hardware models of network nodes and their programming environment, as well as attempts to implement routing algorithms in network node models and their analysis.

Keywords: sensory network · routing algorithms · IoT/IIoT · hazard zones

1 Introduction

Wireless sensor networks are being implemented in mining systems [1], including: pressure sensors, intelligent belt conveyors, telemetry exchange of environmental conditions, monitoring of longwalls and roadheaders. Research and development work is being carried out [2], which envisages the implementation of wireless sensor networks including: pressure sensors, sensor networks to monitor the condition of the rock mass, measurements of the deformation of the support roof, wireless acoustic-optical devices to create escape routes in mines. In such extensive networks, it is necessary to properly organise the transmission of information to ensure operational reliability, flexible configuration and operational safety. It is particularly important to ensure the operational reliability of sensor networks operating in methane and/or coal dust explosion hazards. To

A. Puchalski et al. (Eds.): ICTD 2022, ACM 21, pp. 176–185, 2023.
https://doi.org/10.1007/978-3-031-31719-4_18

ensure mobility and easy implementation of sensor networks, they are in most cases equipped with wireless data transmission interfaces. In addition, advanced transmission algorithms (so-called intelligent routing algorithms) are implemented in the sensor networks, which allow efficient management of the data flow in an extended network infrastructure. The data collected by the sensors is collected, processed and visualised by application platforms. The platforms ensure that the data is properly interpreted, prioritised and connected.

The results of the analysis of individual routing protocols [3], indicate that the protocols based on swarm intelligence (SI) have the highest probability of success for algorithm implementation in sensor-critical networks operating in hazardous environments [9–11, 13–15]. These algorithms are characterised by the fact that, regardless of the number of variables in the solution space, they adapt to the constraints and help to solve the optimisation problem. The use of these algorithms makes it possible to keep the transmission parameters of data transmitted in sensor networks at an appropriate level (e.g. throughput, short data arrival times, high reliability of data arrival at the destination). Sensory networks based on swarm intelligence algorithms are self-organising and multi-redundant. However, it is necessary to implement the algorithms in selected hardware modules in order to verify their effectiveness under real conditions (comparison with computer simulations). The article presents attempts to implement selected routing algorithms in hardware solutions of the sensor network model.

2 Methods

Routing algorithms in sensor networks are used to ensure the optimal path of a sent data packet from the sender to the receiver, with respect to the following criteria [4, 12]:

- route - taking into account the adopted routing metrics,
- simplicity of functioning of the algorithms,
- fault tolerance and stability of operation,
- conversion speed,
- flexibility of routing,
- other application-dependent criteria and available functionality of hardware modules.

Routing algorithms used in sensor networks are divided into proactive, reactive and hybrid [5, 7, 10].

Proactive algorithms store routes between individual network nodes are stored in so-called routing tables. Reactive algorithms, on the other hand, are characterised by route discovery on demand (when a data packet is sent). As a result, there is no need to occupy the memory of individual network nodes that store routing tables. Unfortuanately, the consequence of using such a method is that it takes longer to deliver the data packet to the recipient. Another type of algorithms used in sensor network routing, are hybrid algorithms. In this type of algorithms, selected areas of the network work with proactive algorithms, while their connections are implemented with reactive algorithms. The implementation of the presented algorithms on the nodes of sensory networks, operating in hazardous areas, involves the need to meet several additional requirements, resulting from the construction of individual nodes of the network. The hardware implementation of an individual network node must meet the requirements of the ATEX directive (in particular, the assumption that the radiated power of the transmit/receive antenna must be limited to appropriate values), and the radio system should be a low-energy system (i.e., the nodes have battery power or energy harvesting type).

State-of-the-art research [6, 8, 9] has made it possible to specify detailed criteria for selecting the most favourable path for data packets transmitted in sensor networks (Table 1). The analysis has been carried out in two areas:

- algorithms and networks - throughput, time for the packet to reach the destination, energy savings, reliability of data delivery to the destination,
- hardware - in the course of defining the criteria, the available solutions of sensors and additional components of the system were also taken into account, which must meet the mentioned requirements in the area of algorithms and networks (as a result of the requirements mentioned above) and have an adequate level of reliability of operation of the devices in environments with adverse effects on their performance, safety of operation (especially in the field of energy efficiency of the proposed solution and compliance with ATEX requirements).

The set of criteria adopted (Table 1) is sufficient and covers all the necessary performance characteristics required to be met by the proposed solution.

Table 1. Criteria for selecting the most favourable route of the sent data package.

No.	Criterion [unit]	Criterion description
1	Network throughput [kB/s]	Also referred to as data speed, it is the amount of data transmitted over the network per second
2	Time for the package to reach its destination [ms]	Time measured in milliseconds from the time the data is sent from the source node to the time the data arrives at the destination node
3	Energy savings [mJ/kB]	Amount of energy consumed by a network node during transmission of 1 kB of data
4	Assurance of the delivery of da-ne to the destination [%]	Referred to as PDR (Packet Delivery Ratio), the ratio of packets sent to received

Attempts have been made to implement selected algorithms based on swarm intelligence in hardware: the particle swarm optimisation algorithm, the ant algorithm and the bee algorithm. The following section discusses the implementation of each algorithm in a sensor network node.

2.1 Particle Swarm Optimization Algorithm

The modified particle swarm optimisation algorithm was implemented in the sensor network model as intended.

The algorithm for each particle P_i is defined by the relation:

$$P_i = [X_{i,1}, X_{i,2}, X_{i,3}, \ldots, X_{i,D}] \tag{1}$$

where:

D - the dimension of the particle (in the implementation under consideration, equal to the number of nodes in the sensor network); the particle is placed in a space whose position is determined by the distance from each node of the network,

i - i-th particle,

X - position of the particle in D-th space,

The algorithm calculates speed and position using the following formulae:

$$(t) = w \cdot (t-1) + c_1 \cdot r_1 \cdot (P_{best} - X_i(t-1)) + c_2 \cdot r_2 \cdot (G_{best} - X_i(t-1)) \tag{2}$$

$$(t) = (t-1) + V_i(t) \tag{3}$$

where:

w - parameter that determines the effect of a particle's previous position on its current position,

V_i - the current velocity of the i-th particle,

c_1, c_2- scaling parameters,

r_1, r_2- random numbers,

P_{best} - best-fit local solution to the criterion,

G_{best} - best-fit global solution to the criterion,

t - time.

First, the values of the previous position and velocity and the best-fit local and global solution criteria are determined randomly determined from the interval (0,1 >. Each particle represents a solution, in the case under consideration, the solution is the path from the SOURCE node to the SINK node.

The solution is sought in an optimal way (according to the assumed criteria) using the following objective function for this purpose F_{dest}:

$$F_{dest} = w_1 \cdot RSSI + w_2 \cdot HopToSink \tag{4}$$

where,

w_1, w_2 - weights for each criterion falling within the range (0,1 >,

RSSI - signal power of the radio transmitter of the neighbouring node,

HopToSink - number of jumps (transitions between SOURCE node) to a SINK type node.

The objective function is used to determine the best local solution and the best global solution:

$$P_{best} = \begin{cases} \boldsymbol{P_i}, & if\ (Fdest(P_i) < F_{dest}(P_{best})) \\ \boldsymbol{P_{best}}, & else \end{cases} \tag{5}$$

$$G_{best} = \begin{cases} \boldsymbol{Pi}, & if\ (F_{dest}(P_i) < F_{dest}(G_{best})) \\ \boldsymbol{G_{best}}, & else \end{cases} \tag{6}$$

Once the best-fit criteria have been determined, the position and velocity of the particles are recalculated. This calculation takes place in the SINK node, to which all the information necessary for the processes of subsequent iterations must be sent (each SOURCE node collects information about the state of its neighbours, and then sends it to the SINK node).

With respect to the designated array of neighbours of a node of type SOURCE:

$$N_{wi}\left[w_x, w_y, w_n\right] \tag{7}$$

where:

w_x,w_y,w_n - neighbours of the node w_i,

The index of the best neighbour is determined using the equation:

$$n = X_i \cdot L_N \tag{8}$$

where:

X_i - node location w_i,

L_N - number of neighbours of the node w_i.

Then, knowing the solution, the SINK node sends to the SOURCE nodes information about the selection of the best neighbour.

2.2 Ant Algorithm

The modified ACO (Ant Colony Optimisation) algorithm was implemented in the sensor network model according to the following assumptions.

SOURCE nodes collect information about their neighbours (from FOLLOW_ME-type packets). Data packets (called ants) are sent by SOURCE nodes that leave pheromone in the visited node (incrementing the corresponding variable) and "wander" to a neighbour (which they choose based on a right-likelihood function). The probability of node **j** being selected, by an ant at node **i**, is given by the formula:

$$p_{ij} = \frac{\tau^{\alpha} * \varphi^{\beta}}{\sum_{c_{i,l}} \tau^{\alpha} * \varphi^{\beta}} \tag{9}$$

where:

τ - the amount of pheromone left at each node,

ϕ - value of the local criterion function,

l - number of neighbours of the node i,

α, β - coefficients to select the weighting between the pheromone and the criterion function.

The value of the local criterion function is:

$$\varphi = w1 \cdot RSSI + w2 \cdot 1Htosink + w3 \cdot B_{state} \tag{10}$$

where:

RSSI - radio signal strength of the neighbour node,

H_{to_sink} - number of jumps to the SINK node of a neighbour node,

B_{state} - the state of the battery in the neighbour's node,

w_1, w_2, w_3 - weights of particular criteria.

The criterion function depends on the value of the transmitter power of the neighbouring node, expressed as the value of the RSSI parameter, the number of hops to the SINK node and the battery level of the neighbouring node.

In the case under consideration, the algorithm tries to find paths from the SOURCE node to the SINK node, so that their travel time is as short as possible. It also selects nodes with higher battery levels are selected (as neighbours).

When an agent's (i.e. ant's) packet reaches a SINK-type node, its path is evaluated (based on a comparison with other agents' paths). Among all the solutions, the best one is chosen. After a predetermined number of iterations (number of agents sent), the SINK-type node distributes information about the selection of the best path of the data packet to all SOURCE nodes.

The node remembers the selected neighbour. The ant repeats the selection procedure until it reaches a SINK node. The SINK-type node sends out information about the arrival of the packet (ant) at its destination, which was itself. This procedure is repeated for a fixed number of iterations and is also carried out during the occurrence of certain events (type: loss of a node, a new node in the network, etc.).

2.3 Bee Algorithm

The Bee algorithm has been applied as follows. Three types of data packets BeeScout, BeeForager and BeeDancer are distributed in the network. According to the bee algorithm, scout bees are randomly sent to each node in the network, the scout bees remember the distance travelled and collect information about the energy level of each node in the network, the number of jumps to the SOURCE node. When the scout bees reach their destination (which is a SINK-type node), dancer bees are sent from that node to inform all nodes of the path travelled by the scout bee.

In a SOURCE node, the path sent by the dancer bee is evaluated according to Eq. 11 in the form of:

$$P = \frac{w_1 * \sum RSSI + w_2 * \sum B_{state}}{L} \tag{11}$$

where:

L - path length,

$\sum RSSI$ – sum of the power of the radio transmitters of the nodes included in the path,

$\sum B_{state}$ – sum of the battery state of the nodes included in the pathway.

If the value of the criteria of the travelled path is greater than that of the current best path, then it becomes the best path for the SOURCE node, which will be travelled by the next forager type bee.

Forager bees transmit data from a source node of type SOURCE, to a destination node of type SINK (by the route specified by dancer bees).

3 Results

At this stage of the research, the algorithms have been implemented in a three-node sensor network model according to the assumptions presented. The individual nodes had to transfer packets between each other and send diagnostic data to the PC via a serial link.

It was necessary to develop a way of measuring certain criteria:

- time taken for the package to reach its destination,
- throughput,
- certainty of data delivery to the destination,
- energy consumption of a single network node.

The validation environment consisted of radio modules connected to a PC, configured as shown in Fig. 1.

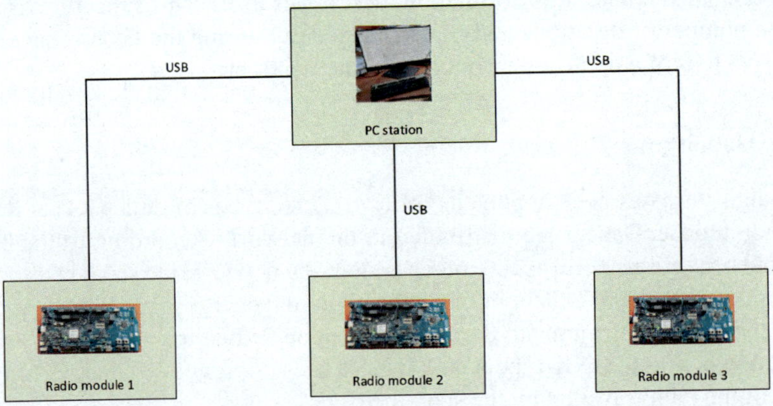

Fig. 1. Validation process test bench.

In order to measure the energy consumed by individual nodes in the network, the current consumed by a single node in the network was measured. The measurement described were performed using the Power Kit Profiler NRF6707 current measurement module, which performs single node current measurements through the NRF Connect application.

During the validation of the algorithms, the following results were obtained by measuring the values of each criterion (Fig. 2).

The value of each criterion was similar for all implemented algorithms. The value of each criterion was similar for all implemented algorithms. The average current consumed by individual network nodes for each algorithm was approximately 1 mA. The network nodes operating with the modified bee algorithm had the lowest energy consumption. The average time to reach the destination node was less than 100 ms, for most packets it was below the value of 50 ms (the orange lines in the graphs in Fig. 2 show the averaged values of the measured time and throughput parameters). The shortest times for a packet to reach its destination were obtained by an algorithm built on the ant algorithm and

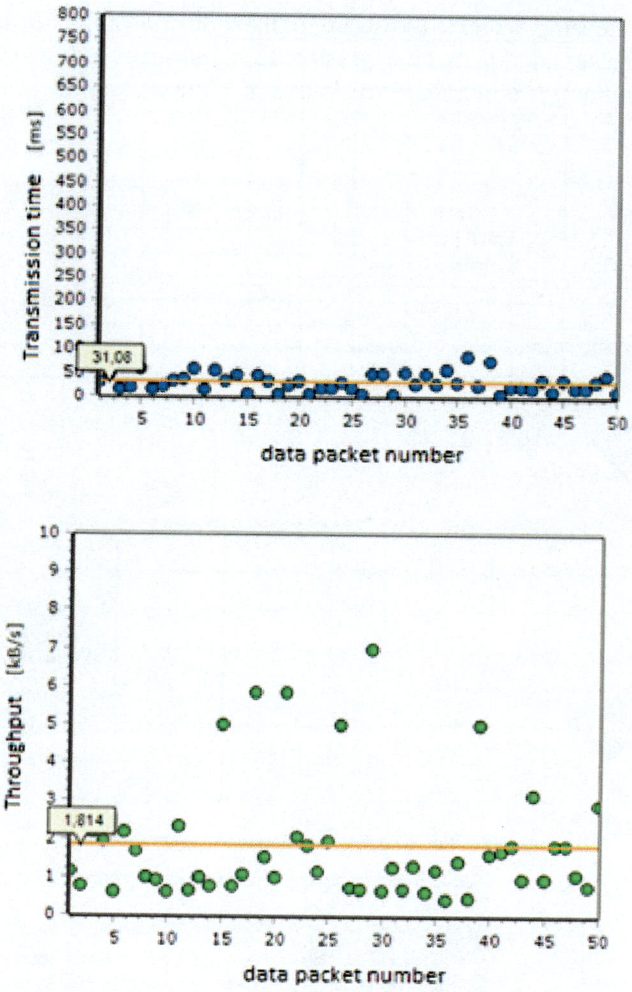

Fig. 2. Packet arrival time to destination and network throughput - data for a single packet transmitted from a NODE node to a SINK node

the particle swarm optimisation algorithm. In the next stages of the research, it is also planned to verify the network reconfiguration time after a node replacement process or node failure. The parameter of certainty of data delivery to the destination for all algorithms was 100% (no packet sent was lost). The purpose of the first phase of the research was to verify the implementability of selected algorithms based on swarm intelligence.

As a result of the research, a testbed configuration concept has also been proposed, which will allow further phases of research to be carried out on a larger number of nodes (Fig. 3). An additional component of the system will be an industrial computer (it is proposed to use a RaspberryPi computer), which will collect data from 4 nodes of the network via UART (Universal Asynchronous Receiver/Transmitter) bus and transmit it to the master unit via Ethernet link (using an Ethernet cable or WiFi network). It is also necessary to develop dedicated software for the industrial computer and modify the software of the master unit (so that it is possible to support the reading of measurement data via Ethernet bus). It is proposed to synchronise the measurement data using the Network Time Protocol (NTP).

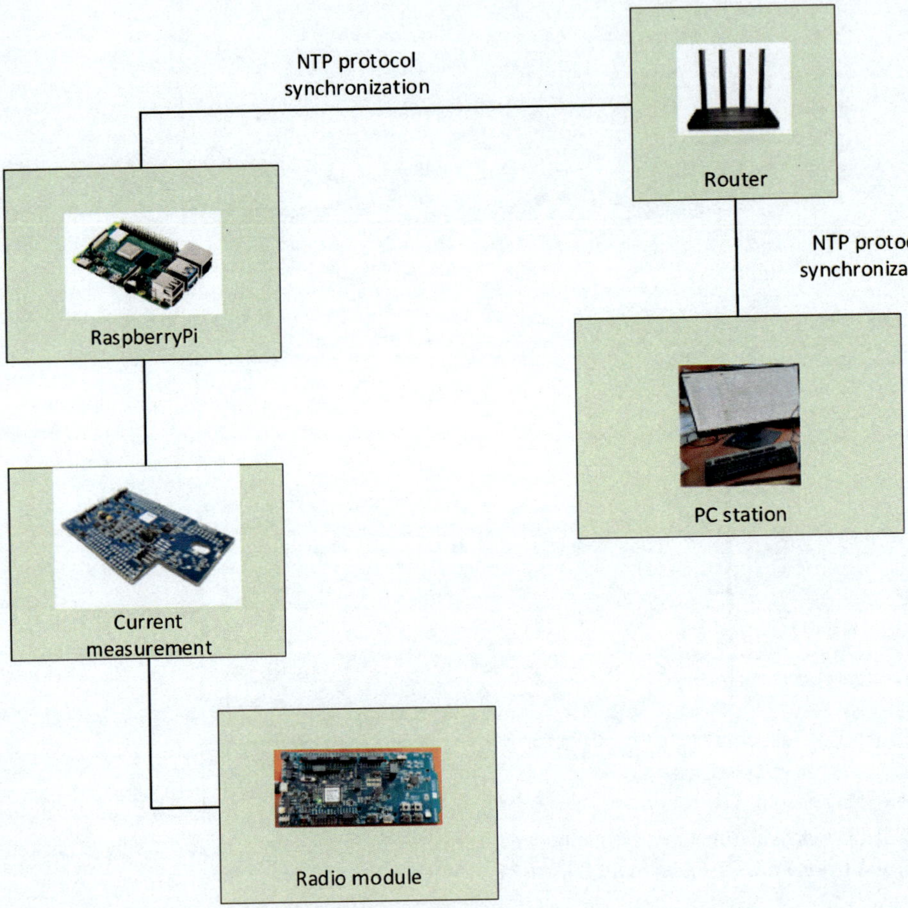

Fig. 3. Proposed test bench for more nodes.

4 Conclusions

This article presents the concept of a routing algorithm for a sensor network operating in methane and/or coal dust hazardous areas.

The concept is based on the use of algorithms based on swarm intelligence to manage the operation of the network. Criteria have been defined for selecting the best path for the transmitted data packet. Three routing algorithms are specified, the process of their implementation in hardware modules of network nodes is described. The implementation process is presented and the conclusions of the implementation are discussed.

It is planned that further stages of research will be carried out on a sensor network consisting of, at least, 25 nodes. It is planned to implement in the network an algorithm based on the ant algorithm using elements of the logic of the particle swarm optimization algorithm.

References

1. Jasiulek, D., Bartoszek, S., Perůtka, K., Korshunov, A., Jagoda, J., Płonka, M.: Shield support monitoring system - operation during the support setting. Acta Montan. Slovaca 2019 **24**(4), 391–401, ISSN 1335–1788
2. Trzop, K., Szurgacz, D., Zhironkin, S., Kasprusz, A.: Analysis of the impact of mining and geological conditions in longwalls on the introduction of pressure monitoring for powered roof supports. IOP Conf. Ser. Mater. Sci. Eng. **1134**(1), 012003 (2021)
3. Zungeru, A.M., Ang, L.-M., Seng, K.P.: Classical and swarm intelligence based routing protocols for wireless sensor networks. A survey and comparison. J. Netw. Comput. Appl. **35**(5), 1508–1536 (2012)
4. Krupanek, B., Bogacz, R.: Routing protocols for Internet of Things systems - simulation studies of selected protocols. Comput. Sci. Electrotechnical Rev. (2018)
5. Jagoda, J., Hetmańczyk, M., Stankiewicz, K.: Dispersed, self-organizing sensory networks supporting the technological processes. Min. Mach. **2**, 13–23 (2021). https://doi.org/10.32056/KOMAG2021.2.2. ISSN 2719–3306
6. Stankiewicz, K., Jagoda, J., Tonkins, M.: Intelligent algorithms for routing sensory networks operating in explosion hazard zones. Min. Sci. **28**, 103–115 (2021). https://doi.org/10.37190/MSC212808. ISSN 2300–9586
7. Lalar, S., Yadav, A.K.: Comparative study of routing protocols in MANET. Orient. J. Comput. Sci. Technol. **10**(1), 174–179 (2017)
8. Gianni A., Di Caro, Ducatelle, F., Gamberdella, L.M.: Swarm intelligence for routing in mobile ad hoc networks. IEEE Swarm Intelligence Symposium 2005. https://doi.org/10.1109/SIS.2005.1501605
9. Kennedy, J., Eberhard, R.: Particle swarm optimization. Materiały IEEE Int. Conf. Neural Netw. **4**, 1942–1948 (1995)
10. Wedde, F.: BeeAdHoc: an energy efficient routing algorithm for mobile ad hoc networks inspired by bee behavior. In: Conference: Genetic and Evolutionary Computation Conference, GECCO 2005, Proceedings, Washington DC, USA, 25–29 June 2005
11. Chan, F.T.S, Tiwari, M.K.: Swarm Intelligence, Focus on Ant and Particle Swarm Optimization, I-Tech Education and Publishing (2007)
12. Smolarek, A., Malinowski, T.: Routing protocols in ad hoc networks. In: Teleinformatics Systems and Networks Conference 2012. Proceedings, pp. 47–60 WWSI 2012
13. Praveena, K.S., Bhargavi, K., Yogeshwari, K.R.: Comparison of PSO algorithm and genetic algorithm in WSN using NS-2. In: 2017 International Conference on Current Trends in Computer, Electrical, Electronics and Communication (CTCEEC), Mysore, India, pp. 513–516 (2017). https://doi.org/10.1109/CTCEEC.2017.8455121
14. Latiff, N.M.A., Tsimenidis, C.C., Sharif, B.S.: Energy-aware clustering for wireless sensor networks using particle swarm optimization. In: 2007 IEEE 18th International Symposium on Personal, Indoor and Mobile Radio Communications, Athens, Greece, pp. 1–5 (2007). https://doi.org/10.1109/PIMRC.2007.4394521
15. Kharati, E.: Using PSO and ABC routing algorithms reducing consumed energy in underwater wireless sensor networks. Iran. J. Sci. Technol. Trans. Electr. Eng. **46**, 367–380 (2022). https://doi.org/10.1007/s40998-021-00474-1

Diagnostics of Worst-Case Execution Time Calculations for Automotive Embedded Systems

Kamil Sternal(✉)

Institute of Fundamentals of Machinery Design, Silesian University of Technology, Gliwice, Poland
ksternal@polsl.pl

Abstract. Estimating Worst-Case Execution Time (WCET) is one of the most important activities in development of automotive embedded systems performing safety critical functions. The importance of WCET estimates comes from the fact that it is one of the inputs used during setting the system's scheduler configuration. Any failure at this stage may lead to a situation where human life could be in danger. Applying a safety factor cannot be a solution, since increasing hardware performance is generating a significant increase in cost of the final product. Therefore, different WCET evaluation methodologies were developed, which is perceived as nontrivial task due to the high complexity of automotive embedded systems and multiple dependencies within structure elements. This paper explains how it is impacting WCET estimation and how WCET can be diagnosed using modern techniques. Depending on the diagnosed system type and the development stage of the product. All these efforts in diagnostics of those automotive systems have in goal to ensure that car is safe under any conditions.

Keyword: Technical diagnostics in automotive engineering

1 Introduction

The Worst-Case Execution Time (WCET) it is a value that can characterize a program or procedure. At the simplest definition [1], WCET is the longest time that will take to execute machine code on a specific hardware platform. Ability to know or estimate that time at a given context allows confirming if time constrains specified for logical function, which is realized by program, are fulfilled. Especially if developed system has to be compliant with ISO 26262 which is automotive standard for functional safety. In Chap. 7.4.17 of ISO 26262 it states that an upper estimation of required resources for the embedded software shall be made, including the execution time [2]. To add more context for how important WCET is in automotive industry, an analysis of technical paper, which is published by one of car manufactures, was made [3]. In total, a referenced vehicle model can have up to 67 different electronic control units (ECU).

A. Puchalski et al. (Eds.): ICTD 2022, ACM 21, pp. 186–193, 2023.
https://doi.org/10.1007/978-3-031-31719-4_19

Nearly one third is realizing safety features like braking, airbags deployment or high-voltage system management. Very often, one safety feature is allocated to multiple ECU's. For example, high-voltage insulation measurement is performed by external sensor but the decision if high-voltage relays shall execute emergency open procedure is made on high-voltage battery control unit (Fig. 1). The insulation measurement, state evaluation, transmission, reception, relay control. All those sub functions are performed by different logical blocks of software. Any unexpected delay can lead to a situation where human life is at risk. Therefore, knowledge abut WCET is so crucial for safety related systems.

This paper is explaining in details what is WCET, what are main difficulties in estimation process and what methods are currently used for that purpose. It is a good starting point for all those who are new into WCET topic.

2 WCET in Details

2.1 WCET in Software Implementation Scope

WCET of software is the maximum time it takes to execute a given piece of code, in a given application context (inputs, state), on a given machine. Figure 2 visualizes how inputs are influencing execution path. For a certain input value, a decision is made and in consequence one or more output path can be executed. As a result, different code blocks will be processed by the microcontroller core and the key characteristic of those blocks is execution time, which combined, represents the majority of WCET value. In addition, execution time for each of the individual blocks is influenced by the platform the code is running.

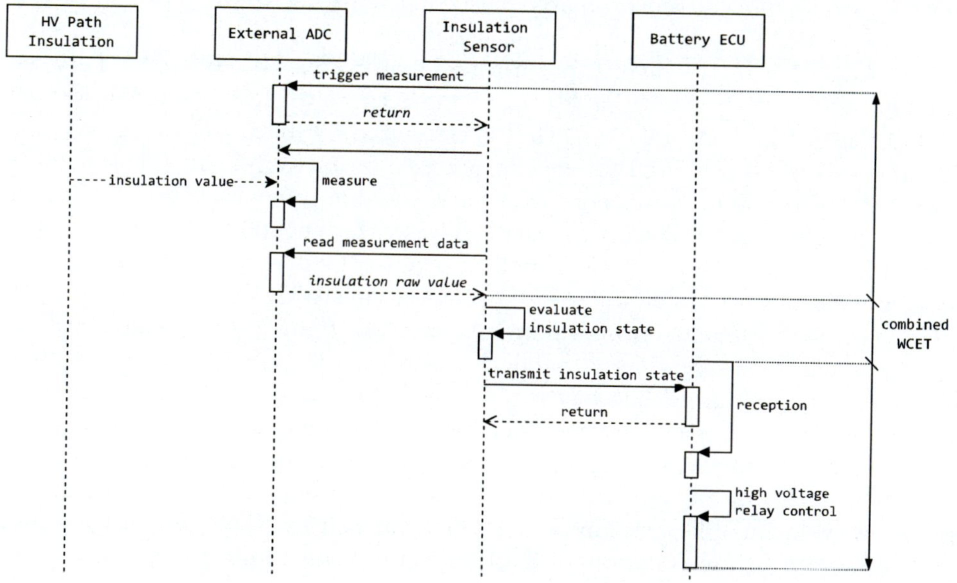

Fig. 1. Example of WCET in context of insulation measurement in automotive system

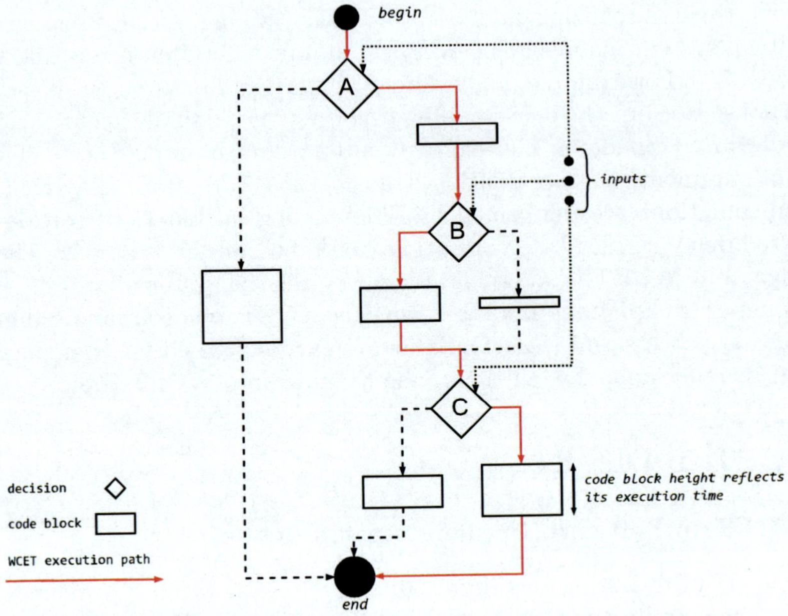

Fig. 2. Visualization of WCET for a logical functionality

2.2 WCET in Software Measurement Scope

Figure 3 is visualizing results after series of executions of an algorithm presented in Fig. 2. The solid line represents the distribution of theoretical measurements. Data of that kind can be evaluated based on static code analyze-one of few methods for WCET estimation. The dashed line is drawn based on data collected during execution of the algorithm on target hardware. The difference between two kinds of data sets is noticeable and that is because the first one is considering best (BCET) and worst (WCET) that is estimated by finding the longest (time criteria) execution path and second one is based on measurements. The point is that this path may be never possible to execute due to characteristics of input data. Strongly coupled data, this means data with dependencies to each other, like air temperature and absolute humidity. With the drop of temperature, also total amount of water in air will drop. It may be that WCET is located in a path where the system reacts on high humidity and low-temperature values. Still, it is always safer to take WCET as validation border. Judgment if observed system is fulfilling time requirements just by considering measured execution times will be as good as are test scenarios. The fact is that with increase in software complexity, creating and execution test scenarios consumes more time because of increased execution paths count. It is worth mentioning that executing the same program on different kinds of architectures, like RISC or CISC, would influence plots on Fig. 3 mainly in distribution of time domain.

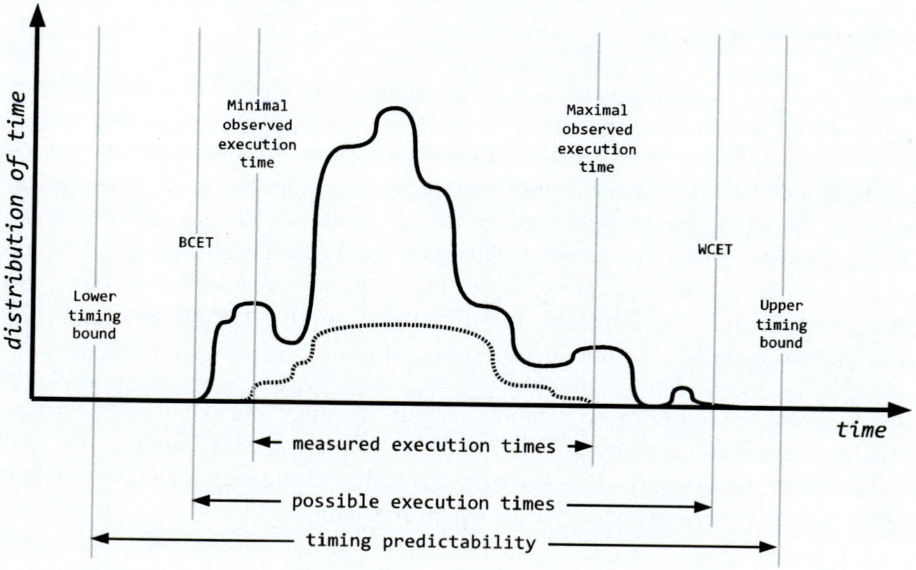

Fig. 3. The WCET/BCET is the longest/shortest execution time possible for a program [1]

2.3 Difficulties in WCET Estimation

Each feature, of an examined system, which is introducing variable execution time within a code block is increasing effort in WCET estimation. This statement is valid for both domains, hardware and software. Figure 4 is visualizing how variable execution time is impacting WCET.

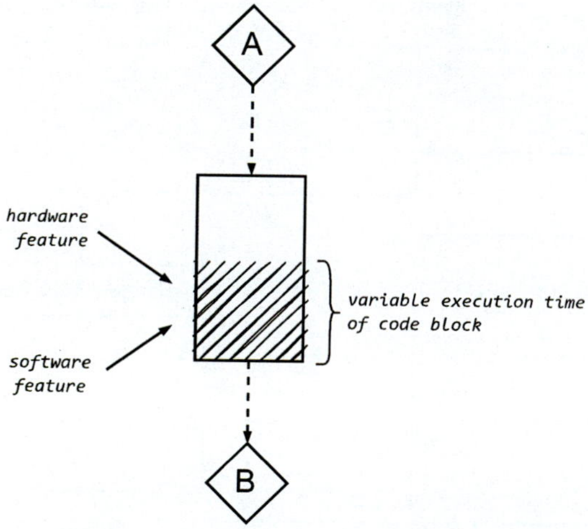

Fig. 4. Feature impact on WCET estimation in software context.

Hardware Features

Memory cache - to increase performance, microcontrollers are equipped with memory cache mechanism. Often used data is copied from slower memory, to faster memory called cache. As a result, the logical core of the microcontroller can fetch data quicker and execute more instructions in shorter time. The decision which data shall be loaded is made based on multiple factors and sometimes leads to situation when unnecessary data is loaded, occupying space for needed data by the current set of instructions. This behavior results in variable code block execution time. In addition, it is difficult to predict the efficiency of cache due to limited technical details in implementation.

Instruction cache - similar to memory cache, an instruction cache shortening instruction execution time by preloading from slower to faster memory. Depending on context, microcontroller registers state, branching etc. for the same code block it is possible to record different execution times.

Interrupts - inextricably bonded with microcontrollers world, are allowing to imminently react on events generated by different peripherals. In systems where interrupts are used extensively, an impact on WCET can be significant. Since interrupt can appear at any moment and can extend the duration of the code block execution by the time needed by the interrupt to be handled [6].

Multicore synchronization - systems where a microcontroller is equipped with more than one core, which is capable of executing user code, are called multicore systems [10]. Such design allows parallelization of code execution, which results in increased computation speed. Such solution enforces use of synchronization

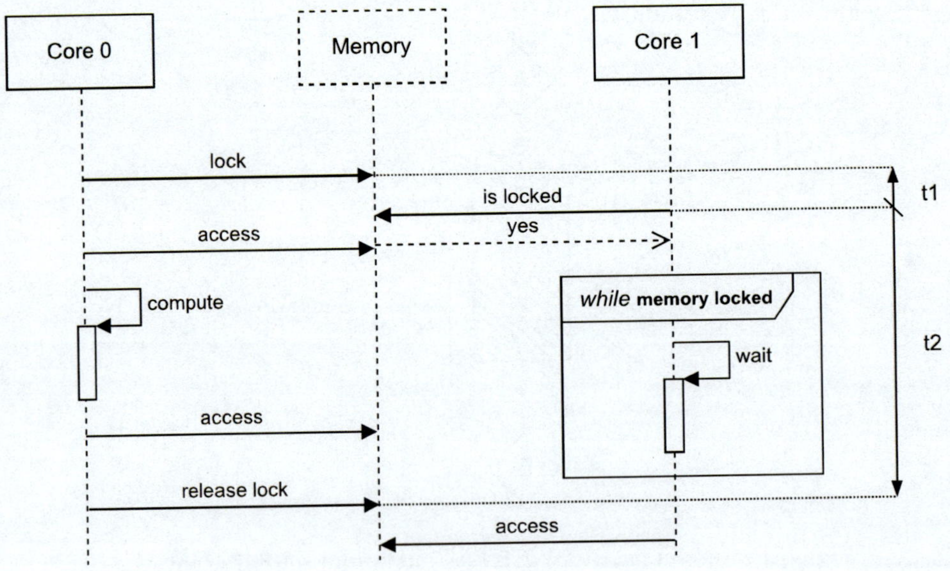

Fig. 5. Memory access on multicore systems

mechanisms. Each time one core would need to exchange data with the other core, it would need to lock a certain memory area. This means that the second core shall not access it while it is marked as locked. In the most common scenario, core two will wait until core one will release lock. WCET for the core two will be influenced by *t1* and *t2*, both marked on Fig. 5.

Software Features

Loop boundary - loop with dynamic boundary, in other words, a loop which repeat count can vary between consecutive calls is affecting WCET. It is worth pointing that a scenario in which loop boundary is depended on external signal, for example user input, can have hidden impact on WCET.

Recursion - in computer science when a function is calling itself then this behavior is called recursion [7]. This affect WCET since a code block that is using recursion will extend it execution time with every single recursive call. If condition to brake recursive call is dynamic, this will result in difficult to estimate WCET. A representation of a code block utilizing recursive call is shown in Fig. 6

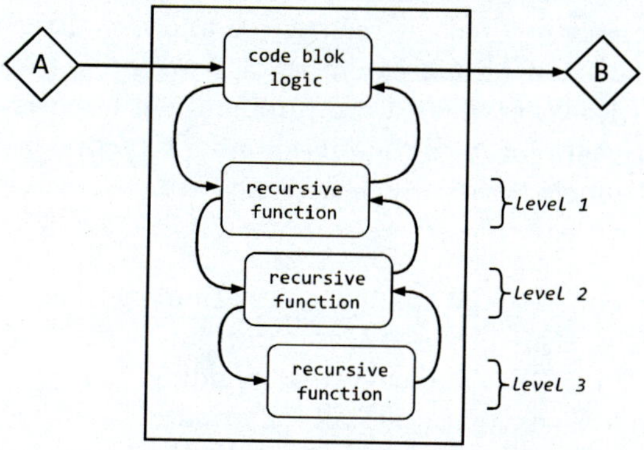

Fig. 6. Code block with embedded recursive call with three levels of deep

Function as parameter - in C or C++ programming languages, from author experience are the most popular for automotive embedded platforms, there is the possibility to pass a pointer to a function. It is powerful mechanism that allow to build robust and highly configurable software mechanisms, but it increases difficulties in WCET estimation by introducing an element (external function) to the software code block which adds variable execution time.

2.4 WCET Estimation Methods

Before WCET methods will be listed and described, it is important to formulate some quality measures that are:

- WCET bounds must be **safe** - must never underestimate the WCET
- WCET bounds should be **tight** - must not be too pessimistic
- The analysis cost should be **reasonable** - analysis is not too resource-intensive

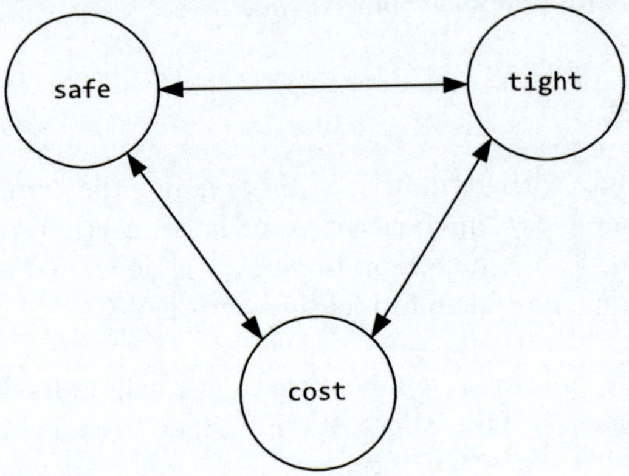

Fig. 7. WCET estimation process properties

Statistical method - it is based on measurement data from target hardware. Multiple test scenarios are executed on the system and execution time is collected. Usually, the granularity of measurement data is set on software function level. This means that each executed software function will be instrumented. In some cases, when resources are limited, the measurement's scope is narrowed to entry functions of logical functionality. Such collected time data together sequence of sampled software entities are used to estimate WCET for depicted scenarios [8].

Static method - it can also be named simulation method. As input, it takes source code, which is then processed and decomposed. Based on decomposed object code, a flow analysis is made [11]. Depending on how sophisticated is the method, it is possible to extract statically allowed flows or actual feasible paths which are giving safest and tightest results with the biggest estimation cost. To get WCET this method in addition needs detailed knowledge about the target platform. This is required for translation of object code into time blocks, which then can be summed based on earlier estimated paths.

Hybrid method - it is a combination of statistical and static methods. Measurements are used to estimate WCET of code blocks, but paths are evaluated based on static analysis of code. This approach eliminates disadvantages of previous methods. Test scenarios can be replaced by routines which main goal is to call every available function in the tested project. This is a safer approach since it is easy to verify if all functions were instrumented. In case of wrong designed or missing test scenario, the WCET could be wrongly estimated. Moreover, the main disadvantage of static methods is eliminated need of detailed knowledge about internal mechanics of target platform. Since all timing data doesn't have to be estimated but easily measured [9].

3 Summary

WCET estimation problematic is well understood and described in literature. That facts are giving this topic a potential for high level of automatization. Yet, from the author professional experience, this potential is not fully utilized during development of commercial products. The statistical method is the most preferred one, which can be implemented on existing testing environments used during development. Remaining methods are requiring skilled engineers, dedicated test environment and time for building initial setup and fine-tuning. The author predicts that rapid development of machine learning techniques can make hybrid method more popular in the near future.

References

1. Gliwa, P.: Embedded Software Timing. Springer International Publishing (2021)
2. ISO: ISO 26262 road vehicles - functional safety. ISO, Geneva, Switzerland (2011)
3. Audi: Self Study Program 684, Audi e-tron GT, Audi, Germany, 130–140 (2021)
4. Ranawat, D., Prasad, M.P.R., Srinivasan, G.: An efficient approach for wcet measurement of runnables and tasks of the application software: autosar compliant multi-core partitioning. In: 2019 IEEE 1st International Conference on Energy, Systems and Information Processing (ICESIP). IEEE (2019)
5. Bai, Y.: Practical microcontroller engineering with ARM technology. John Wiley and Sons, Nashville, TN, 373 (2016)
6. Ganssle, J.G., Barr, M.: Embedded Systems Dictionary. CMP Books, London, England, 19–20 (2003)
7. Knowlton, T.: Programming in C++. Course Technology, Florence, AL, 438 (1996)
8. Piroddi, L.: Special Topics in Information Technology. Springer International Publishing (2022)
9. Cazorla, F.J., Kosmidis, L., Mezzetti, E., Hernandez, C., Abella, J., Vardanega, T.: Probabilistic Worst-Case Timing Analysis. ACM Comput. Surv. **52** (2019)
10. Domeika, M.: Software development for embedded multi-core systems. Newnes, London, England (2008)
11. Lokuciejewski, P., Marwedel, P.: Worst-case execution time aware compilation techniques for real-time systems. Springer, Dordrecht, Netherlands, 11 (2010). https://doi.org/10.1007/978-90-481-9929-7

Application of Torque Signal Analysis of Servomotors to Assess of Support System Condition of Industrial Machining Centre

Damian Augustyn(✉) and Marek Fidali

Silesian University of Technology, Gliwice, Poland
{damian.augustyn,marek.fidali}@polsl.pl

Abstract. Supports systems of machining centers can cause many maintenance and process problems. Improper support system operation is caused, among other things, by wear of mechanical parts and faults in drives. To detect faulty support system operation, measurements of the torque of its servomotors can be used. Servomotor controllers can measure motor current, which can be easily used to calculate motor torque. For research purposes, the authors used a set of torque signals from machining center servomotors acquired over a long period. The signals were collected during a diagnostic test programmed into the machining center controller and performed once a day. In this article, a method of analyzing torque signals for assessing the condition of machining center support systems is presented. During the research, statistical as well as custom-developed torque signal estimators were considered, which were used to develop a diagnostic model of the support system representing its good condition. The assessment was based on the output of a neural (ANN) model. The results show that a simple ANN model is a sufficient solution for detecting faults in the investigated support systems and can easily be applied in a condition monitoring system.

Keywords: Condition monitoring · Torque signal · Machining centre · ANN

1 Introduction

Contemporary machining centers used in production are usually fully automated and complex mechatronic systems. They are equipped with a number of supports, depending on the complexity of the center and machining process. An example of a support system for a multi-axis machine is presented in Fig. 1.

Reliable operation of the support system depends on the conditions of each of the system components, as presented in Fig. 2.

The reliable operation of support systems in machining centers is crucial for the smooth and efficient functioning of the machining process. Common faults that occur in support systems include problems with the servo motors, increased resistance in movement, decay of preload, increased clearance, ball screw damage, and bearing damage [1, 2]. As machining processes take place in a confined space and harsh conditions, direct diagnosis of support system faults can be challenging. Nevertheless, the condition of the support systems can be evaluated through the use of indirect diagnostic methods.

A. Puchalski et al. (Eds.): ICTD 2022, ACM 21, pp. 194–202, 2023.
https://doi.org/10.1007/978-3-031-31719-4_20

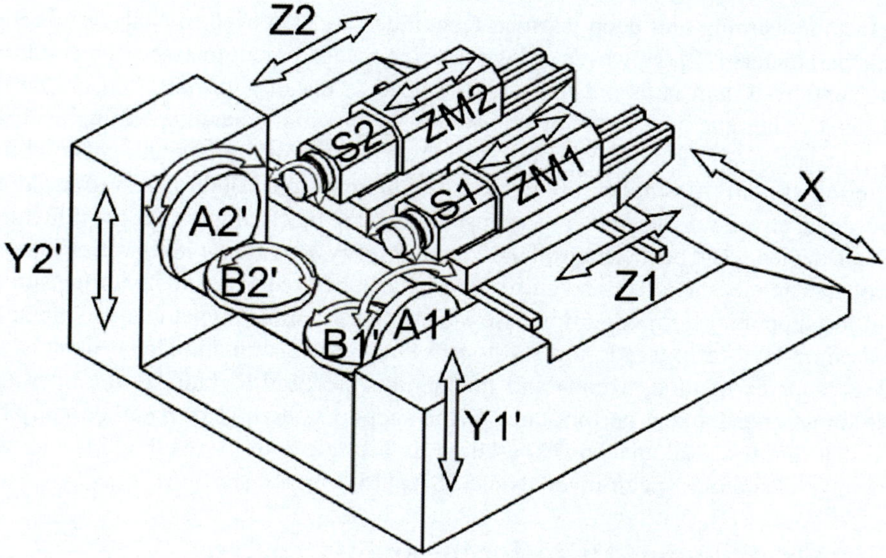

Fig. 1. Diagram of the axis support systems of a double-spindle machining centre

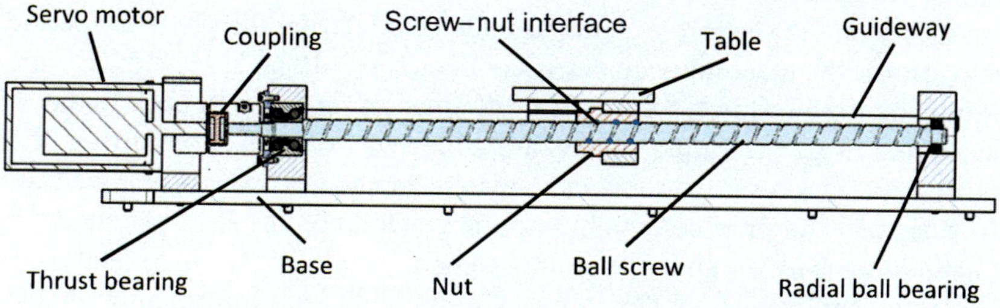

Fig. 2. Schematic diagram of the ball-screw drive system

1.1 Condition Monitoring and Diagnostics of Support Systems

Support systems in machining centers can be monitored and diagnosed in various ways. Indirect assessments are often based on the evaluation of machining part quality [3]. Additionally, there is the possibility of measuring vibration, as reported in [4–12]. Methods for vibration measurement of the support system include statistical analysis, spectral and time-frequency analysis, and other approaches [8–11]. The condition of the support system can also be evaluated using numerical, experimental, and operational modal analysis methods [4–7]. Modal analysis enables the estimation of a modal model, including damping, natural frequencies, and mode shapes, during the operation of the machining center and enables monitoring of its condition by comparing modal models for a new support system and one estimated for normal operation. This can reveal faulty operation of the support system.

Machine learning and deep learning techniques are also used to analyze vibration signals parameters [12]. However, using vibration measurement to assess the condition of the support system is associated with the need to install additional sensors on the machining center and a high risk of sensor damage due to the harsh operating environment. The use of additional sensors also increases the cost of implementing a condition monitoring system. A simpler way to assess support system conditions is to use diagnostic signals from sensors and devices installed by the machining center manufacturer. One such device could be the controller of the support system's servomotors. Servomotor controllers can measure electrical current, which can be used to calculate torque signals useful for support system condition assessment. Servo torque signal analysis methods are based on long-term trends and short-term fluctuations, utilizing signal models and the least squares method. Trends and fluctuations are used to indicate the operating condition and mechanical performance of the support system [13]. Another study has found that torque signal analysis is based on multi-scale entropy (MSE) [14] and, less frequently, on singular spectrum analysis (SSA) [15].

2 Signals of Support Drive Torque and Its Analysis

During our research, we analyzed torque signals collected by our industrial partner from December 2019 to September 2022. The torque was measured during diagnostic tests on the drives of the X, Y, Z, and A axes of 47 machines. According to the maintenance service, no faults in the support systems were detected during the observed period. The diagnostic test for the machining center is conducted regularly, at least once a day, and involves moving the axes drives through a full range of motion and back to the starting position. During this test, torque signals are collected from the servo drives of the X, Y, Z, and A axes. Figures 3a-d show examples of the torque signals obtained for each axis. These signals can provide insight into the general condition of the entire support system, including the drives and support mechanisms. The signals can be analyzed in the amplitude, time, and frequency domains, and with additional information about speed and distance, spatial analysis can also be performed. A simple statistical analysis is initially performed, but more advanced signal analysis methods can be used for further research.

During the research, a set of 100,000 signals was processed and analyzed. Before analysis, the signals were validated and preprocessed. This was necessary because the signals differed in the number of time points and the timing of the beginning of the diagnostic test. During the validation, the signals were matched and corrupted ones were rejected.

The preprocessed signals were subjected to statistical analysis. Initially, 16-point estimators were used on the whole signal and its selected parts. The features used in the feature vector were selected based on correlation analysis, which allowed the determination of relevant features and rejection of redundant ones. The analysis reduced the number of features to 10. The identifiers of features and their description are presented in Table 1.

Further analysis showed that the selected features were the best to determine the condition of the support systems. Relevant features of the torque signals for the X-axis drive are presented in Fig. 4.

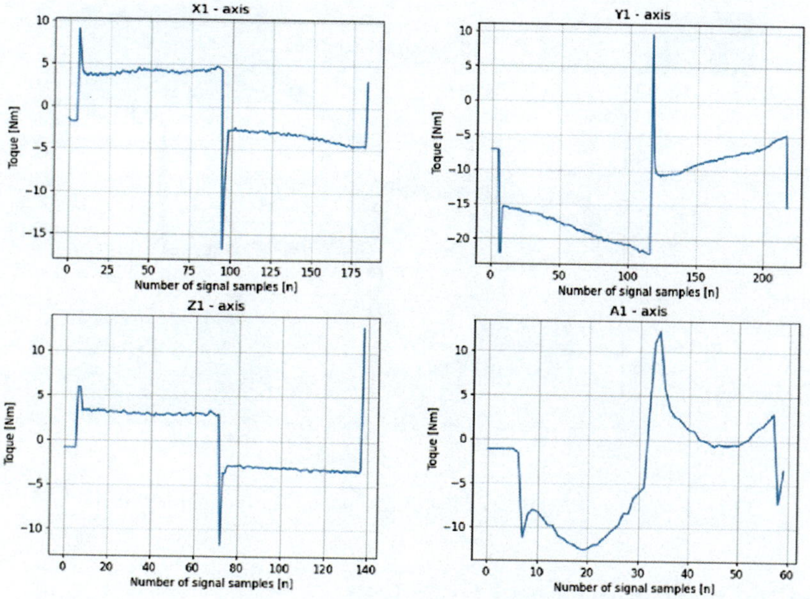

Fig. 3. Torque signal of the support servo motors for the individual axes

Table 1. Feature identifiers with a description

Feature vector	Description of the feature
Max_p1	Maximum value determined from area P1 (Fig. 4)
Mean_p2	Mean value determined from area P2 (Fig. 4)
Var_p2	Variance value determined from area P2 (Fig. 4)
Kurt_p2	Kurtosis value determined from area P2 (Fig. 4)
Max_p2	Maximum value determined from area P2 (Fig. 4)
Min_p3	Minimum value determined from area P3 (Fig. 4)
Mean_p4	Mean value determined from area P4 (Fig. 4)
Var_p4	Variance value determined from area P4 (Fig. 4)
Kurt_p4	Kurtosis value determined from area P4 (Fig. 4)
Min_p4	Minimum value determined from area P4 (Fig. 4)

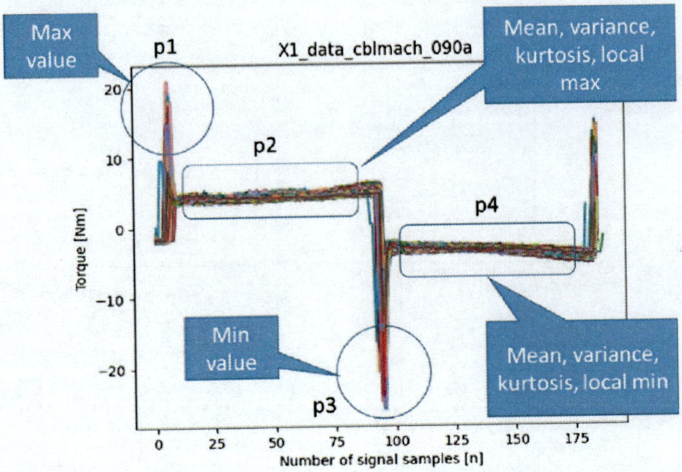

Fig. 4. Selected features of the torque signal for the feature vector

3 Method of Support Fault Detection

Taking into account the results of the torque signals analysis and the method of their acquisition, we propose a method for assessing the general condition of the machining center's support system. The general idea of the method is depicted in Fig. 5. The torque signals of the individual axes' servo drives, measured during periodic diagnostic tests, are stored in a database. The recorded data is then processed and analyzed. Based on the relevant feature vector, which is the result of the torque signals analysis, the condition of support system of each axis is classified. We assume that the support system classifier is trained on the feature values of torque signals that represent supports in good condition. The performance of the classifier can be considered as a condition indicator and tracked over time. In the case of support system degradation, we expect that the condition indicator will decrease. Using a lower limit control method, we will be able to detect changes in support condition and alert maintenance services that an additional inspection is necessary.

Different classifiers can be used in the method. At the beginning of our research, we decided to use a neural classifier based on a classical backpropagation artificial neural network (ANN). In the case of the developed method, the response of the ANN network is considered as the classifier performance.

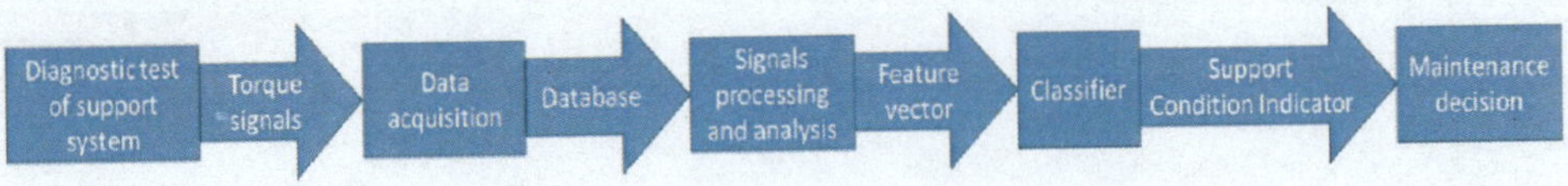

Fig. 5. Method of diagnosing support system

3.1 Data Preparation

Feature values estimated from torque signals were used for classification of support condition. To prepare the training and testing sets, it was necessary to process the feature values. As all available data represented good support conditions, the feature values were ordered in descending order to find and discard outliers. 25% of the lowest and the highest values were treated as outliers with probable anomalies and were discarded from the classifier training. The remaining 50% of feature values were assumed to be connected with torque signals describing good support conditions and were used for classifier training. Outliers were used for classifier testing. The method of selecting feature values is presented in Fig. 6. Additionally, to test the classifier, a vector of feature values was prepared to simulate hypothetical faulty operation of the support.

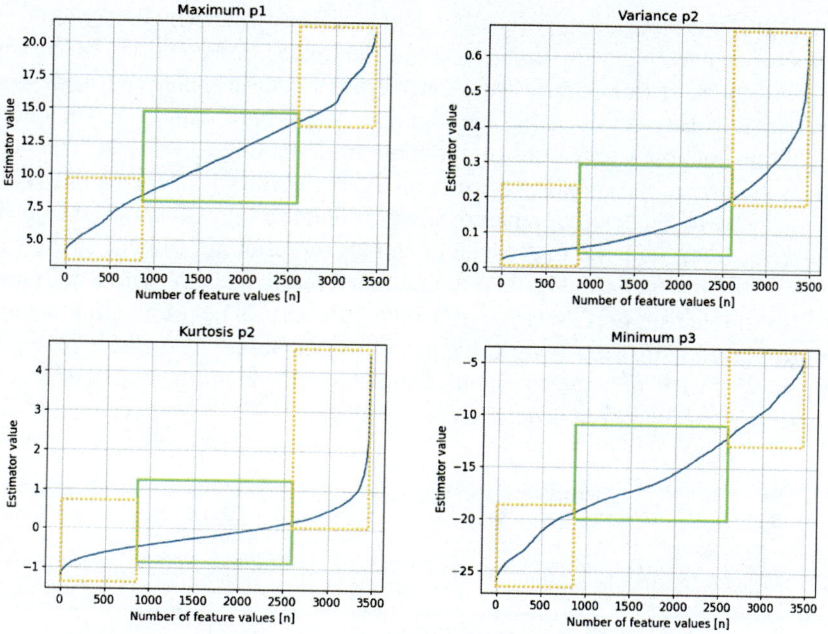

Fig. 6. Ordered from smallest to largest selected relevant feature values. The marked areas show the method of selecting features for neural classifier training (green box) and testing (yellow boxes).

3.2 An ANN Classifier

As previously mentioned, a classifier based on a classical feed-forward artificial neural network was created for classification purposes. A Python environment with the Tensorflow and Keras libraries was used for computation. The ANN network has three layers: input, one hidden, and output. Ten input neurons were created for the input layer, corresponding to the length of the developed feature vector consisting of 10 features. The hidden layer consists of 16 neurons and uses a sigmoid activation function. The output layer of the network consists of one neuron, which displays the status indicator of the machining center's drive supports. The network was trained on the prepared training data, which contained data representing the good condition of the machining center drive and outliers considered uncertain. The Adam optimization method was used to train the network, which is known for its high optimization efficiency. The developed neural network achieved a classification accuracy of over 98% on the prepared validation data.

4 Classification Results and Discussion

After training the ANN classifier, it was tested on data that was not presented to the classifier during training. The developed network classifies the set of feature vectors representing the good condition of the machining center drives with an average efficiency of 99%. For the set of feature vectors that differ from good state, the average efficiency is 76%.

Table 2 presents an exemplary result of the classifier testing. The first column shows the results for an exemplary feature vector representing a good condition of the support system. The values of this feature were taken from a dataset where the ANN was not present. The ANN network classifies the support system status with 99% efficiency. The second column contains the feature vector taken from the outliers dataset. The network determines the state of the support system with an efficiency of 82%. In the third column, the feature vector has specially prepared values of features that simulate damage. These values were hand-crafted due to the lack of signals representing damage to the support systems and differ more strongly than the Outlier vector. The ANN network classifies the conditions of the support system with a 69% efficiency. The fourth column contains the feature vector with each feature having the same value. The result of the ANN network shows that it recognizes significant differences from the signal representing good condition and determines the efficiency at 0%.

Table 2. ANN classification results

Feature vector	Correct vector	Outlier vector	Anomaly vector	Constant vector
Max_p1	12.1439	17.0579	18.5542	1
Mean_p2	4.5994	5.2612	6.3949	1
Var_p2	0.0785	0.3781	0.6231	1
Kurt_p2	0.1360	−0.0235	−0.1826	1
Max_p2	5.3740	6.2287	6.0373	1
Min_p3	−13.4430	−13.2793	−23.8638	1
Mean_p4	−3.8064	−4.3654	−5.1618	1
Var_p4	0.1281	0.1483	0.5247	1
Kurt_p4	−0.4731	−0.4455	−1.0485	1
Min_p4	−4.4863	−5.2574	−5.3245	1
ANN	0.999909	0.824204	0.696919	0.003550

5 Conclusions

The use of servo drive torque signal measurement is an interesting alternative to assess the condition of a machining center support system. In this article, a method is presented that allows for the general assessment of the support system's condition based on a classifier trained on data representing good conditions of the support system. As a training dataset, statistical features of torque signals were considered. These features were estimated during the torque signal analysis process. An ANN was used as the classifier. The ANN was trained based on the torque signals' relevant feature vectors. The ANN successfully recognized examples deviating from good conditions, confirming that this approach could be used for the general assessment of the support system's condition.

References

1. Kim, Y., Bae, H., Kim, S., Vachtsevanos, G.: Fault diagnosis of AC servo motor with current signals based on wavelet decomposition and template matching methods. IFAC Proc. Volumes **41**, 7239–7244 (2008)
2. Lee, W.G., et al.: Failure diagnosis system for a ball-screw by using vibration signals. Shock Vibr. **2015**(9), 1–9 (2015)
3. Archenti, A., Laspas, T.: Accuracy and performance analysis of machine tools. In: Gao, W. (ed.) Metrology. PM, pp. 215–244. Springer, Singapore (2019). https://doi.org/10.1007/978-981-10-4938-5_7
4. Choi, Y.H., Cha, S.M., Hong, J.H., Choi, J.H.: A study on the vibration analysis of a ball screw feed drive system. Mater. Sci. Forum **471–472**, 149–154 (2004)
5. Jia, P., Rong, Y. Huang, Y.: Condition monitoring of the feed drive system of a machine tool based on long-term operational modal analysis. Int. J. Mach. Tools Manuf. **146**(6), 103454 (2019)
6. Li, K., et al.: Vibration-based health monitoring of ball screw in changing operational conditions. J. Manuf. Process. **53**, 55–68 (2020)
7. Ellinger, J., et al.: Feed drive condition monitoring using modal parameters. MM Sci. J. **2019**, 3206–3213 (2019)
8. Nayana, B.R., Geethanjali, P.: Analysis of statistical time-domain features effectiveness in identification of bearing faults from vibration signal. IEEE Sens. J. **17**, 5618–5625 (2017)
9. Moussa, M.A., Boucherma, M., Khezzar, A.: A detection method for induction motor bar fault using sidelobes leakage phenomenon of the sliding discrete fourier transform. IEEE Trans. Power Electron. **32**, 5560–5572 (2017)
10. Seshadrinath, J., Singh, B., Panigrahi, B.K.: Investigation of vibration signatures for multiple fault diagnosis in variable frequency drives using complex wavelets. IEEE Trans. Power Electron. **29**, 936–945 (2014)
11. Ben Ali, J., Fnaiech, N., Saidi, L., Chebel-Morello, B. Fnaiech, F.: Application of empirical mode decomposition and artificial neural network for automatic bearing fault diagnosis based on vibration signals. Appl. Acoust. **89**, 16–27 (2015)
12. Jiang, G., He, H., Xie, P., Tang, Y.: Stacked multilevel-denoising autoencoders: a new representation learning approach for wind turbine gearbox fault diagnosis. IEEE Trans. Instrum. Meas. **66**, 2391–2402 (2017)
13. Zhang, X., Zhang, Q., Tan, L. Xu, G.: Running state detection and performance evaluation method for feed mechanism of numerical control machine. In: 2017 IEEE International Conference on Prognostics and Health Management, ICPHM 2017, pp. 222–226 (2017)

14. Chang, J.L., Chao, J.A., Huang, Y.C., Chen, J.S.: Prognostic experiment for ball screw preload loss of machine tool through the Hilbert-Huang transform and multiscale entropy method. In: 2010 IEEE International Conference on Information and Automation, ICIA 2010, pp. 376–380 (2010)
15. Xu, X., Zhao, M., Lin, J.: Detecting weak position fluctuations from encoder signal using singular spectrum analysis. ISA Trans. **71**, 440–447 (2017)
16. Jais, I.K.M., Ismail, A.R., Nisa, S.Q.: Adam optimization algorithm for wide and deep neural network. Knowl. Eng. Data Sci. **2**, 41 (2019)

Fuzzy Condition Indicator of Spindle of Machining Center Based on Vibration Parameters

Damian Augustyn(✉) and Marek Fidali

Silesian University of Technology, Gliwice, Poland
{damian.augustyn,marek.fidali}@polsl.pl

Abstract. Vibration condition monitoring of machining centre spindles are most often used. Many machining centres are equipped with vibration continuous monitoring systems which are able to detect not only spindle faults (like bearing defects) but also problems occurring during machining process. The authors are focused on spindles condition assessment based on vibration parameters. The purpose of the research was developing of fuzzy logic system to classify the general condition of the machining centres spindles based on vibration severity presented in ISO standards. As a result of fuzzy system is general condition indicator which could be simply trended and assessed by maintenance personnel in order to detect a spindle problems during it operation. For the testing of the systems we used real life data acquired during production process in machining centre over 3 years. Results shows that such approach could be useful for evaluation of machining centre spindle condition changes over machine life and additionally could be used in automated inference system in the future.

Keywords: Condition indicator · Vibration parameters · Spindle · Fuzzy logic

1 Introduction

Machine centers are very complex systems generating many signals which could be used for the condition monitoring of different subsystems. The most important parts of machining centers are spindles. Machining centers could have more than one spindle which operates simultaneously with different speeds and tools.

Manufactures of machining centers equips them in different monitoring systems which allows detects severe faults and stop the process to avoid costly damages [1–4]. Vibration monitoring systems are the most often used in machining centers to diagnose spindles conditions and detect toll collisions [5–8]. There are different diagnostic parameters of spindle vibration signals which could be monitored. To assess general spindle conditions an effective (RMS) vibration velocity is controlled and compared with vibration severity levels recommended in the ISO standards. There are also parameters estimated for high frequency filtered acceleration signals. The commonly known are 0-P and RMS and allow to assess bearing conditions.

A. Puchalski et al. (Eds.): ICTD 2022, ACM 21, pp. 203–211, 2023.
https://doi.org/10.1007/978-3-031-31719-4_21

Despite of details of vibration monitoring systems there are two main problems in real life with using them. The first one is that the systems are part of a machine manufacturer maintenance system with access only for manufacturer maintenance personnel and internal machine monitoring system. The second problem is that plant maintenance services are not always aware that such solutions exist. But even they know that monitoring system exists and is possible to get access to the data the problem is in its interpretation, understanding and assessment of spindle condition.

2 Vibration of Spindles and Assessment of its Condition

Vibration condition monitoring of machining centre spindles are most often used. Many machining centres are equipped with vibration continuous monitoring systems which are able to detect not only spindle faults (like bearing defects) but also problems occurring during machining process. However, the monitoring of vibrations and the observation of their trend change does not have to be related to the progressive wear of the machining centre spindle. The increase in the average vibration value may be significantly influenced by changes introduced in the machining parameters, such as an increase in the spindle speed. The figure below (Fig. 1) shows the RMS vibration velocity (vRMS) in frequency band 10-1000Hz and corresponding spindle rotational speed time series.

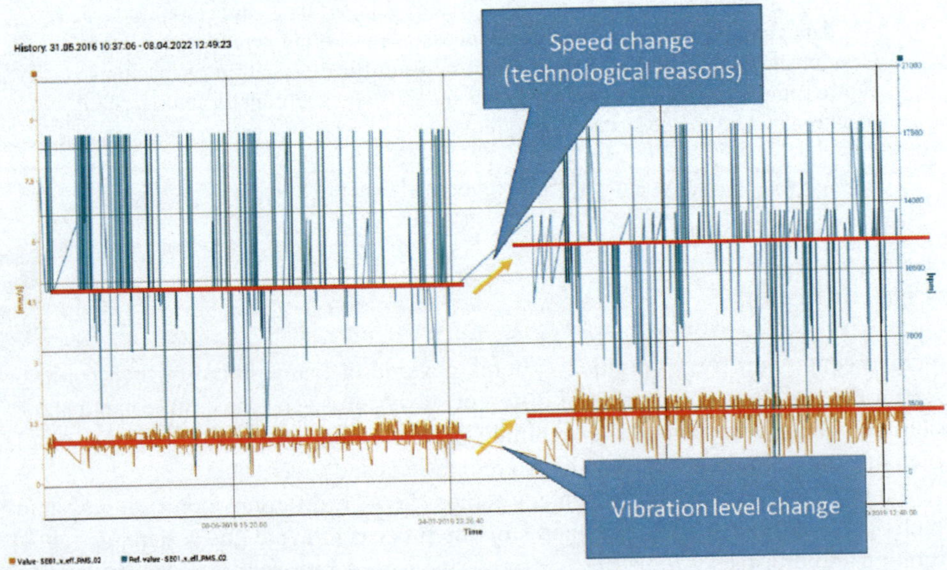

Fig. 1. Presentation of the impact of changing the spindle speed on the average vibration value

When analyzing this graph, it can be noticed that the trend does not have to be related to the damage progress, but to technological changes. This example shows that the vibration amplitude changes as the speed value changes. Increasing the machining speed from week to week also increases the average vibration level. The relationship between speed and vibration is strong especially when speed values are close or covers to natural frequencies of spindle-support system. The change in the average vibration value of a time series must not only depend on the change in the spindle speed of the machining center. The type of mounted tool, changes in cutting parameters or changes in the material parameters of the workpiece can also have a large impact on vibration level. Each of these factors should be taken into account when building a fuzzy inference model. Authors focused on vibration and speed value because those data were only available. In the Fig. 2 are presented the time series of vibration RMS velocity values of the three double spindle machining centers. For each spindle, the observed vibration trend is increasing, however, without additional information about the process parameters, we cannot say what cause that vibration grow. For the presented machines, the warning state and the critical state were determined on the basis of fixed limit values. Setting the fixed limits for the spindle which operating conditions are variable may raise doubts. This practice may lead to false alarms on the one hand, and the risk of undetected faults on the other hand.

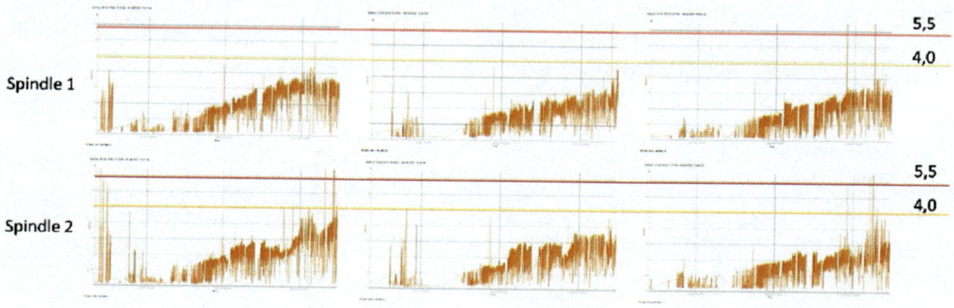

Fig. 2. Analysis of RMS vibration velocity of time series and set limits

2.1 Spindle Condition Assessment Using Standards

To assess the condition of the spindles we can use one of the two ISO standards. The first is a series of ISO 17243 standards dedicated to the machine tool spindle. It assumes the measurement of the RMS velocity of vibrations in the frequency band 35-3500Hz. The standard recommends the use of fixed limits. We can also use the classic ISO 10816 - 3 standard (ISO 20816) dedicated to the assessment of the condition of typical industrial machines. The standard also assumes the measurement of the RMS velocity of vibrations, while in the narrower frequency band 10-1000Hz. The ISO 10816 standard also assumes a constant value of the limit thresholds. Both of the described standards allow for the assessment of the condition of the machining center spindle, but their criteria do not take into account the process variability. In modern machining centers, process variability plays an important role, because vibrations change very dynamically during one machining cycle. Often, simultaneous machining is performed on the two or more spindles, which results in vibration interactions causing changes in the vibration level. Limits can be adjusted to the actual vibration values but there will be a problem of fixed limit values with the risk of possible false alarms.

Taking into consideration tendency to automation of diagnostic inference in computerized spindle condition monitoring systems, it could be possible to apply a solution based on artificial intelligence to solve the problem of spindle general condition assessment. The various AI techniques like artificial neural networks (ANNs), fuzzy logic, fuzzy neural networks (FNs), genetic algorithms etc. could be used in this solution [9–14] – we decide to use a fuzzy logic system to create a general spindle condition indicator.

3 General Indicator of Spindle Condition Based on Fuzzy Logic

Fuzzy logics are multivalued logics that form a suitable basic for system reasoning under uncertainty or vagueness that allows intermediate values to be defined between conventional evaluations like true/false, yes/no, high/low [15]. For designing the spindle condition monitoring system we use vibration values collected automatically by existing in machining centre measurement and data collection system. Additionally, we considered also rotational speed for which measurements of the spindle vibration values were made. The following illustration (Fig. 3) shows a diagram of general spindle condition indicator estimation based on fuzzy logic.

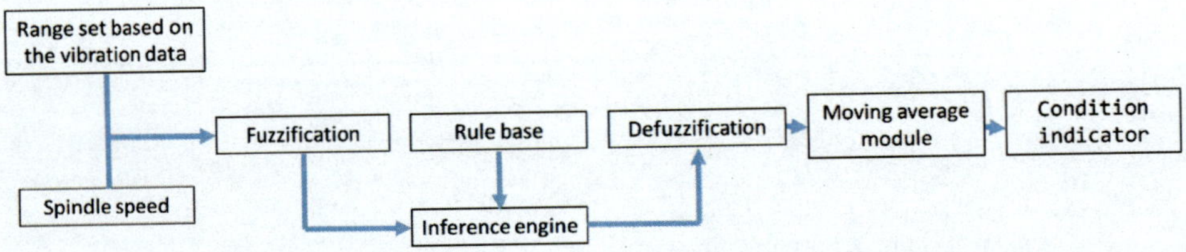

Fig. 3. Scheme of operation of the developed program of the general indicator of the spindle condition based on fuzzy logic

The system is based on two types of fuzzy sets represented RMS vibration level and spindle rotational speed (Fig. 4), for which five membership functions were proposed: very low, low, medium, high and very high. The first of the sets concerns the vibration values. The ranges of membership functions and fuzzy rules were based on the ISO standards vibration severity at the preliminary stage of the research. Finally we decide to design membership functions and fuzzy rules based on analysis of the acquired vibration and rotational speed values. The second set concerns the values of the spindle rotational speed, where also own determined fuzzy parameters where used. On the basis of the developed fuzzy sets, a fuzzification of the vibration and rotational speed values were performed. Then the inference engine using the developed rule base (Table 1) for the fuzzy input generates the fuzzy output, which undergoes the process of defuzzification in order to transform the output into a crisp value. The output membership function has three values: NC (Normal condition), WC (Warning condition), and EC (Emergency condition).

The fuzzy inference system was based on the Mamdani method [16]. At the end of the system, a moving average module has been added for smoothing of the value of the overall spindle condition indicator of the machining center.

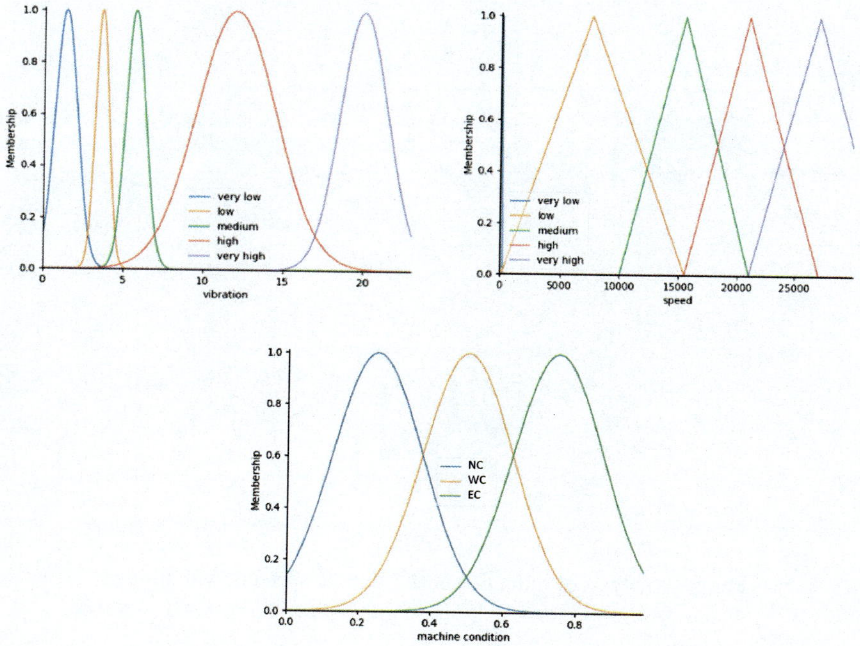

Fig. 4. Fuzzy sets created for the developed system

Table 1. Fuzzy rules for spindle condition monitoring system

Vibrations	Very high	EC	EC	EC	EC	EC
	High	EC	EC	WC	EC	EC
	Medium	WC	WC	NC	WC	WC
	Low	NC	NC	NC	NC	NC
	Very low	NC	NC	NC	NC	NC
		Very low	Low	Medium	High	Very high
		Rotational speed				

Relation between vibration and rotational speed are presented on charts in Fig. 5. Analysis of this charts allows us to determine ranges of the membership functions and fuzzy rules. When analyzing the graphs, we can see that the relationship between the vibration and speed values is not linear. The highest values of the vibration amplitude can be seen in the middle of the spindle speed range of the machining center. This nonlinearity is connected with dynamic properties of the spindle support and was taken into consideration during the preparation of fuzzy rules.

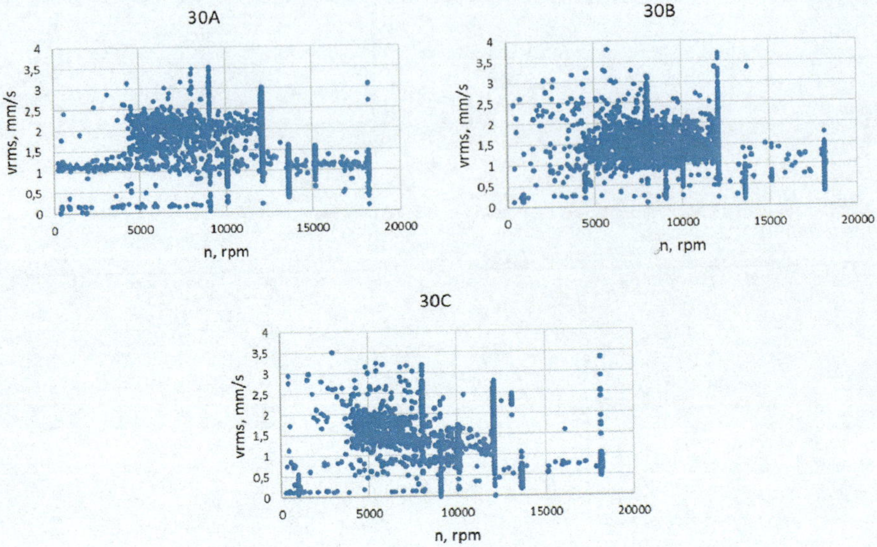

Fig. 5. Analysis of influence of the spindle speed on vibrations

3.1 Results

During the research we tested fuzzy system under different parameters. The Fig. 6 shows a comparison results of the fuzzy system where only ISO 17243 standard vibration severity was used to develop fuzzy sets (left chart) and system based on own determined parameters of vibration and rotational speed (right chart). When analyzing the result of a system based only on ISO vibration severity, we can see a strong deviations of condition indicator value and trend of value change due to variability of the rotational speed. The fuzzy system where the rotational speed values were taken into account shows that there is no significant growth trend of condition indicator values. After the smoothing of the indicator values we can better the effect of vibration speed influence. Tests for six double-spindle machining centers have confirmed the operation of the developed system.

In the Fig. 7 result of the tests performed for one of the machines is presented. It confirms the effectiveness of the developed system, showing the sensitivity of the system to the real increase in vibration value, taking into account the variability of the machining process.

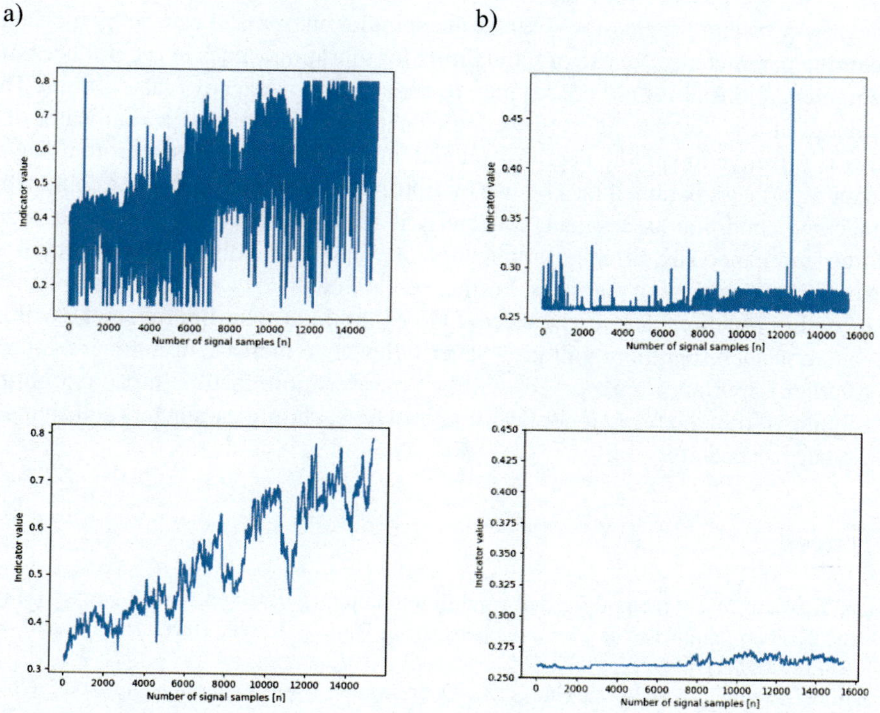

Fig. 6. Comparison of a simple spindle condition indicator system based on ISO standards (left) with the developed system (right)

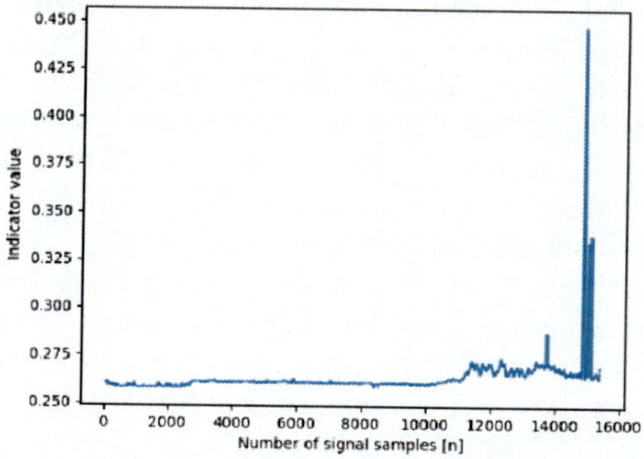

Fig. 7. An example showing the sensitivity of the condition indicator to a real increase in vibration

4 Conclusions

The research results presented in the paper show that it is possible to develop a fuzzy condition indicator of the machining center based on vibration parameters. In order to develop the condition indicator, the recorded vibration signals of the spindles of the machining centers had to be checked. The analysis showed a strong dependence of the vibration values on the speed of the spindle. The upward tendency of the vibration signal does not always mean progressive wear of the spindle, but may be caused by a change in machining parameters. The use of fixed limits for vibration values in accordance with the recommendations of ISO standards may result in the appearance of false alarms. The development of a system based on fuzzy logic enables the creation of an indicator of a state that is sensitive to spindle degradation and, at the same time, resistant to changes in machining process parameters. The use of a proposed condition indicator could simplify machine condition assessment, especially if many machines of similar class are monitored simultaneously. Effective application of proposed condition indicator requires setting of limits for alarming purposes. Further research will be focused on improvement of membership functions and fuzzy rules. One of the important improvement will be taking into consideration more vibration severity levels to increase of number of fuzzy stets. Another important improvement will be consider more factors impacting on the results obtained from the fuzzy model, such as tool type, cutting parameters and changes in workpiece parameters.

References

1. Tian, Y., Wang, W.: A method for tool condition monitoring based on spindle power. Tianjin Daxue Xuebao (Ziran Kexue yu Gongcheng Jishu Ban)/J. Tianjin Univ. Sci. Technol. **54**, 1179–1186 (2021)
2. Janak, L., Stetina, J., Fiala, Z., Hadas, Z.: Quantities and sensors for machine tool spindle condition monitoring. MM Sci. J. **2016**, 1648–1653 (2016)
3. Chen, X., Zhang, X., Cao, H.: Advances in condition monitoring, diagnosis and vibration control of smart spindles. Jixie Gongcheng Xuebao/J. Mech. Eng. **54**, 58–69 (2018)
4. Yan, R., Gao, R.X., Zhang, L., Lee, K.B.: A module-based software system for spindle condition monitoring. Int. J. Mechatron. Manuf. Syst. **2**, 532–551 (2009)
5. Žarnovský, J., Kováč, I., Mikuš, R., Fries, J., Mošat', M.: Vibration diagnostics of CNC machining center spindle. Manuf. Technol. **19**, 350–356 (2019)
6. Cizikova, A., Monkova, K., Monka, P., Urban, M.: Analysis of frequency characteristics at spindle of CNC machining centre. MM Sci. J. **2016**, 1515–1518 (2016)
7. Vogl, G.W., Donmez, M.A.: A defect-driven diagnostic method for machine tool spindles. CIRP Ann. Manuf. Technol. **64**, 377–380 (2015)
8. Zhou, Z.D., Chen, Y.P., Fuh, J.Y.H., Nee, A.Y.C.: Integrated condition monitoring and fault diagnosis for modern manufacturing systems. CIRP Ann. Manuf. Technol. **49**, 387–390 (2000)
9. Shantisagar, K., Jegadeeshwaran, R., Sakthivel, G., Alamelu Manghai, T.M.: Vibration based tool insert health monitoring using decision tree and fuzzy logic. SDHM Struct. Durability Health Monit. **13**, 303–316 (2019)
10. Deore, K.S. Khandekar, Mrs. M.A.: Design machine condition monitoring system for ISO 10816-3 standard using fuzzy logic. Int. J. Eng. Res. Technol. (IJERT) **4**, 726–729 (2015)
11. Jayaswal, P., Verma, S.N., Wadhwani, A.K.: Application of ANN, fuzzy logic and wavelet transform in machine fault diagnosis using vibration signal analysis. J. Qual. Maint. Eng. **16**, 190–213 (2010)
12. Frieß, U., Kolouch, M., Friedrich, A., Zander, A.: Fuzzy-clustering of machine states for condition monitoring. CIRP J. Manuf. Sci. Technol. **23**, 64–77 (2018)
13. Zhang, W., Jia, M.-P., Zhu, L., Yan, X.-A.: Comprehensive overview on computational intelligence techniques for machinery condition monitoring and fault diagnosis. Chin. J. Mech. Eng. **30**(4), 782–795 (2017). https://doi.org/10.1007/s10033-017-0150-0

14. Mushiri, T., Mbohwa, C.: Vibration based condition monitoring of rotating machinery using fuzzy logic. In: Lecture Notes in Engineering and Computer Science vol. 2226, pp. 1018–1022. Newswood Limited (2016)
15. Zadeh, L.A.: Is there a need for fuzzy logic? Inf. Sci. **178**, 2751–2779 (2008)
16. Izquierdo, S.S., Izquierdo, L.R.: Mamdani fuzzy systems for modelling and simulation: a critical assessment. J. Artif. Soc. Soc. Simul. **21**(3), 2 (2018)

Check for updates

Angular Backlashes Monitoring in Heavy Industrial Machines

Pavlo Krot[1(✉)], Volodymyr Korennoi[2], Radosław Zimroz[1], and Jarosław Szrek[3]

[1] Faculty of Geoengineering, Mining and Geology, Wroclaw University of Science and Technology, 50-421 Wroclaw, Poland
{pavlo.krot,radoslaw.zimroz}@pwr.edu.pl

[2] Iron and Steel Institute of Z.I. Nekrasov, National Academy of Sciences of Ukraine, Dnipro 49000, Ukraine
koren_noy@ukr.net

[3] Faculty of Mechanical Engineering, Wroclaw University of Science and Technology, 50-370 Wroclaw, Poland
jaroslaw.szrek@pwr.edu.pl

Abstract. The heavy industrial machines are subjected to intensive wear of critical parts in the drivelines, which resulted in angular backlashes, causing high dynamic loads and abrupt failures. Backlash diagnostics is a complicated task for maintenance personnel. A comprehensive analysis of methods and instrumentation for the angular backlashes measurements in the drivelines of heavy machines in different industries is represented. The advantages and shortcomings have been shown of different principles of wear measurement in the teeth couplings, universal spindles and gearboxes of drivelines with a complex structure. The principle of backlashes measurement by the reversal motion of drives is illustrated by the computer simulation. The digital device for angular backlashes measurement is developed and created. Accuracy is estimated depending on the rotation speed and sampling rate of the signals. Results of industrial testing in the drivelines of industrial machines of different designs are demonstrated. The conditions for further implementation of this method and sensor requirements in the online monitoring systems of industrial plants are discussed.

Keywords: Drivelines · Angular Backlashes · Measurement · Sensors · Condition Monitoring

1 Introduction

Angular and radial clearances in the drivelines of various machines and mechanisms are the inevitable consequence of their operation on transmitting torsional load from the electrical motors or propulsion engine to the working tools. Clearances, or backlashes, while sustained within moderate ranges play a positive role, allowing easy installation of large-sized components and assembly units, compensating possible distortions and displacements, providing lubricant flow on all contacting surfaces and eliminating jamming

A. Puchalski et al. (Eds.): ICTD 2022, ACM 21, pp. 212–228, 2023.
https://doi.org/10.1007/978-3-031-31719-4_22

of parts, for example, due to thermal expansion of bearing parts. However, with wear increasing, gaps begin to play a negative role, especially when the cyclical or reversible load is applied to the working tools of the machine, for example, in the drivelines of rolling mills and mining machines. In precision actuators and robotic manipulators, open gaps lead to positioning errors and control delays. In automotive units, the presence of backlash of the steering bodies and transmission also leads to emergencies during operation. With the increased transmitted torsional loads, the dynamic torques dramatically rise due to the non-linear stiffness of couplings resulting in overloading and failures of machines.

The mining and metallurgical machines, which drivelines consist of large inertial masses like rolls, gears, wheels, couplings and electric motors, work under non-stationary loads and speeds. All elements of the drivelines experience large static and dynamic torques and high specific contact loads, which generate and contribute to the inevitable wear of the contacting surfaces (spindle pads, gear teeth and couplings). For example, wear of 5–6 mm in the spindle bronze pads of rolling stand happens in 1–3 weeks depending on rotation speed and drive load. In its turn, this wear (angular backlash) causes 3–5 times higher dynamic torque in the drivelines.

Therefore, the angular gaps in the driveline and the radial gaps in the supports should be considered the main parameters of the technical condition of the mechanical equipment, affecting the dynamic loads and failure rates. The local defects of gears and bearings are a consequence of increased dynamic loads. The solution to the problem of diagnosing the open part of the total gaps and their total size is a challenging technical and scientific problem due to a lack of suitable sensors.

1.1 Approaches to Equipment Maintenance

According to the practice of industrial rotating machinery operation, the total wear should always be controlled by the maintenance staff as the main parameter of equipment. However, the standards of condition monitoring [1, 2] do not give the tools and methods for their automatic monitoring or manual measurement in the large-scale mechanical parts of metallurgical and mining machines.

The open part of the backlash characterizes the initial settings, operation modes and design features of the drivelines. If the gap is closed at the time of load applied to the working tool and does not open during further oscillations, system dynamics is not affected but gap size cannot be diagnosed in a transient mode.

The maintenance staff for the service actions planning is interested to know the total wear of parts. On the other hand, the gaps open part, which affects the dynamics and is preferable for diagnostics, is not a criterion for the appointment of maintenance actions and parts replacement. Hence, certain contradiction exists between maintenance regulations and requirements to diagnostic methods and tools. From the equipment servicing point of view, total wear should be diagnosed, and from the point of view of the effect on equipment overloads, the open part of the total clearance is more important and should be diagnosed. Resolving this contradiction can have two approaches.

The first approach requires the development of algorithms for technical condition monitoring invariant to the open gap size when estimating the total wear. To go from the open gap size to its total value (wear of parts), functional relations are needed between them, or a dynamic model, which accounts the influence of equipment settings (spindle weight balancing, tightening of bolts) and working parameters, e.g. speed ratio of roll and metal slab, on open part of the gap. This is enough complicated task requiring individual models for unique equipment.

The second approach supposes adaptation of existing regulations and transition to maintenance by the dynamic loads instead of wear. However, this approach requires continuous monitoring of mechanical loads in the equipment, comparing them with the allowable limits, calculating the remaining useful life of the equipment parts and assigning date of replacement [3].

The second approach is not quite similar with standard one based on deep diagnostics of incipient defects (tooth cracks, damage to bearings), which requires vibration measurements and other non-destructive testing methods. However, existing methods of vibration diagnostics at the steady state of equipment operation, e.g. envelope spectrum, are difficult to implement for total wear recognition in the drivelines. When machine is loaded by external torque, gaps are closed, and at idle mode, they are in an uncertain state, i.e. periodically open when the shaft rotates. Currently, there are only a few experimental studies to determine the changing open part of the gaps in the drivelines at idle mode. The completed studies are based on a combination of the two above approaches; the method given further allows diagnosing both the open part of the gaps and total wear.

1.2 Factors of Gaps Play and Torsional Dynamics

In the dynamic analysis of machines, the nonlinearity corresponding to the gap (also called as dead zone) is mathematically described, as a rule, by the various kinds of piecewise linear characteristics, see Fig. 1 where F – force (torque), δ – generalized coordinate (angular or linear displacement). The photo in Fig. 1a is a spindle head coupling with the wearable bronze pads. Figure 1b shows an analytical representation of angular clearance as a dead zone in the driveline caused by wear or contacting bodies. Figure 1c shows "softening" function of bearing support stiffness where fracture point means contact opening in the bolted joints on the gearbox housing. Figure 1d corresponds to approximation of "hardening" function when stiffness increases above certain level of applied load [4].

To analyze the factors influencing gaps play during the operation of machines, extensive measurements were performed in the drivelines of several stands of the hot rolling mill [5]. These studies have shown that the angular backlash opening at the time when load applied to the rolls, i.e. when capturing a strip, depends on such factors:

- spindle rotation speed and head slot position (inertial forces);
- strip linear speed approaching to the rolls (ratio of strip and rolls speed);
- spindles weight balancing (periodical drops in the gaps of bronze slipping pads);
- ratio of the static and dynamic component of the torque (affects gaps opening).

To determine the imbalance of the spindles, as the main reason for opening the gaps, vibration measurements were made on the sliding bearings supports (Fig. 2a) of the upper and lower spindles at idle mode and under load. For the absence of oscillations from the balanced state, the deviations of the forces in the supports should not exceed 4–5% (100–150 kg) of the weight of the spindle. The spring supports of spindles with

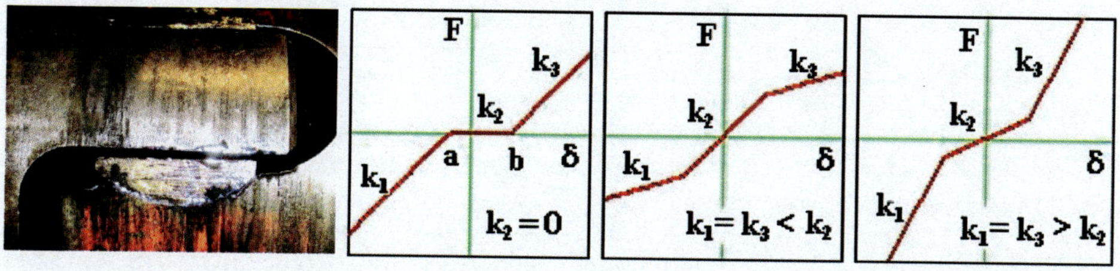

(a) (b) (c) (d)

Fig. 1. Typical piecewise linear stiffness characteristics of the elastic couplings: (a) angular clearance in the spindles; (b), (c), (d) analytical representations by bilinear functions.

screw adjustment clamps used on most rolling stands. They do not allow, in principle, even approximate balancing of the spindles to eliminate oscillations during rotation and the associated opening of the gaps.

The total stiffness of the two springs in supports, for example, in the investigated roughing stands is 560, 840 and 3100 N/mm. Therefore, the installation accuracy of the fixing nuts on the racks of the spindle supports, taking into account their total weight of spindles with four bronze pads and bearing (about 2–3 t), should be no more than 2–5 mm in different stands. In industrial conditions with the existing design of supports (diameter of a regulating screw is 50–60 mm), it is difficult to maintain the necessary tuning accuracy even with the use of vibration meter, according to a method [6]. Hence, in this case and for many other applications, e.g. in transmissions of heavy-duty load-haul-dumping (LHD) vehicles with frequent transient loading [7] or reverse rolling mills with vertical rolls [8], the gaps appear opened due to underbalanced or overbalanced. This leads to a significant increase of the Torque Amplification Factor (TAF):

$$TAF = M_{max}/M_{stat}, \tag{1}$$

where M_{max} – maximum dynamic torque, M_{stat} – static loading torque.

The graph in Fig. 2b shows the dependence of the TAF in the one of drivelines of the rolling mill on the angular position of the spindle head slot when metal piece is captured by the rolls.

Given the low average level of dynamics, deviations of TAF are also not significant (within 10–20%). Maximum dynamical loads are observed when the slot is horizontal. The ratio of static and dynamic loads in a given elastic connection plays a role in opening the gaps in subsequent torsional oscillations after load application. If the dynamic component exceeds the static moment, then gaps will open, and subsequent back shock will occur. If the dynamic torque does not change its sign (no crossing of zero level), then only the initial gaps opened before the load applied will affect the dynamics of the system.

In addition to the angular gaps, there are also radial clearances in the bearing supports of the gearbox shafts and rolls chocks. Their influence is in the redistribution of the contact stresses on the gears causing teeth damages at the end edges during the transient torsional oscillations [9]. To prevent gaps play and reduce torsional dynamics, different methods of electric drive control can be implemented [10–13] in combination with

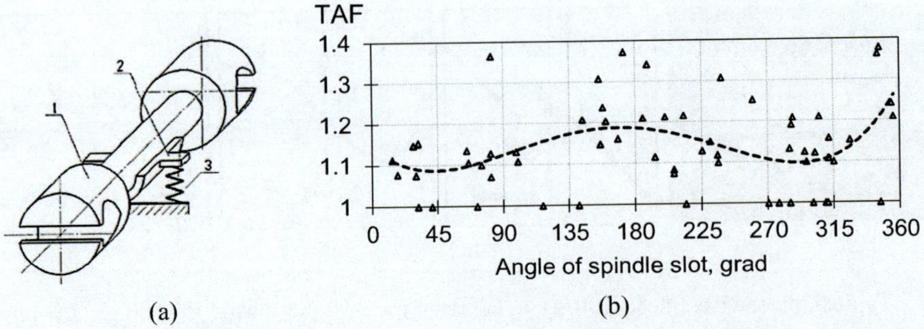

Fig. 2. Design of spring type support of spindle (a): 1 – head; 2 – support; 3 – springs; (b) dependence of TAF on the angular position of spindle head slot: 0 (360), 180° – horizontal; 90, 270° – vertical.

methods of gaps diagnostics by the signals of drive speed and load currents [14, 15]. However, some of these methods are only applicable to gearless drivelines. The presence of reducers requires involvement of dynamical models to estimate backlashes [16–23].

In this paper, authors are mainly focused on the instrumentation and sensors for the direct measurement of the angular clearances.

1.3 Methods and Instrumentation for Backlashes Measurement

Diagnostics of wear (angular gaps) on rotating shafts is an important part of maintenance procedures in many industries:

- automotive (transmissions and steering devices);
- oil and gas (reverse drilling equipment);
- energy (couplings of electric machines and turbine generators);
- shipbuilding (shafting with internal combustion engines);
- aviation (steering mechanisms and servo drives);
- metallurgical machines (drivelines of rolling stands and coilers);
- mining equipment (excavators, long wall shearer, conveyors, loaders).

The use of standard sensors and vibration signals for diagnostics of gaps by known methods allows determining wear only indirectly, by changing the surface geometry of the contacting parts (teeth, rolling elements), or by harmonics of the shaft beat frequency, if radial loads allow it to move in bearing bearings with gaps. In the gearboxes of most large rolling mills, the ratio of the weight of toothed shafts and radial spacer forces is such that movements occur only in transient capture-ejection modes of the workpieces. At idle and under load, it is very difficult to identify the absolute values of the gaps from the vibration signals on the shaft bearings, since radial forces at low speeds of rotation keep the shaft in one of the extreme positions. Use of special proximity sensors on each shaft support, as practiced in turbine units, is problematic in harsh working conditions.

The standard tools and data processing methods for diagnosing gaps with sensors in the drivelines are not yet sufficiently developed. The only method for measuring clearances given in regulatory documents is to measure using thickness-calibrated probes during machines repairs. The mechanical system for backlashes measurement in the large-scale gearboxes is proposed in [24] (see Fig. 3a) where denoted: 1 – gear coupling shaft; 2 – bushing hub; 3 – flexible connection; 4 – spring; 5 – pressing clamp; 6 – intermediate roller; 7 – reducer; 8 – rotation angle sensor; 9 – housing.

The system based on inductive and optical sensors is developed in [25], which scheme is shown in Fig. 3b where denoted: disk 6 with radial slots, aligned with the points of intersection of the teeth 4 involute with the pitch circle (*d*); 7 – optical sensor; 8 – photo detector of the optical sensor; 9 –induction type sensor.

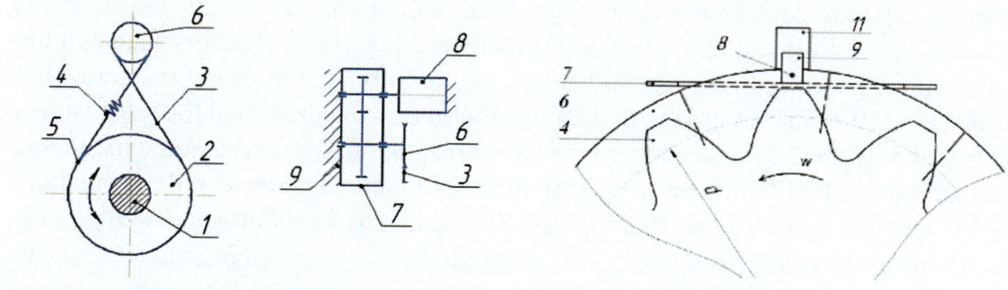

(a) (b)

Fig. 3. Backlashes (wear) measurement systems based on: (a) mechanical encoder [24] (adopted); inductive and optical sensors [25] (adopted).

In automotive diagnostics [26], as well as in other industries, the basic principle of backlash measurement is the static fixation of the initial position of one part of the articulated node and the application of the torque to its second part. The sensors can be either dial gauges or optical sensors similar to those used in laser shaft alignment.

This principle is acceptable for small-sized machines, or if there are appropriate test benches. Under production conditions, it is desirable to perform diagnostics using current technological loads and drives. For this, inertial closure and opening of the gaps in transmissions can be used when reversing or accelerating and decelerating the drive or working body of the machines.

According to the method [27], to determine the wear of the main transmission line of vehicles, the transmission is rotated sequentially in the acceleration and coasting modes. As a parameter characterizing the wear of the gear teeth, take the ratio of the maximum amplitude of the variable component of the axial reaction on the supports of the transmission input shaft in acceleration mode to the average value of this variable component in coasting mode. The difference in signals is due to the shorter transmission time of one of the sides of the teeth. Obviously, this method is not suitable for reverse gearboxes where both sides wear is almost identical.

Quite a lot of patent sources devoted to the measurement of gaps in screw mechanisms. For example, a patent [28] provides a device for determining wear in friction pairs of a rolling stand pressure device (screw thread and screw thrust bearing) in the course of mill operation. The device contains sensors of the angular position of the screws, sensors for moving the screws within their stroke, as well as sensors for the position of the height of the roll necks connected via toothed rails on the beam of the weight-balancing device. The signals are memorized, compared with previous values and thus the wear is determined.

In the [29], an example of diagnostics of the friction unit shaft of an aircraft landing gear hinge is given, the tested node is affected by a compressive force, the compression deformation is measured and the characteristic of the node stiffness is built. The transition from one section to another is characterized by a fracture characteristic, the position of which determines the size of the gap in the friction unit. By constructing the characteristics of stiffness in various directions of action, we obtain the values of wear at various points of contact of the elements of the friction unit.

In work [30], in order to diagnose the wear value of the bed linings and work roll pillows, an experimental assessment of the structural damping in the finishing stands of a wide-strip hot rolling mill 2500 was performed based on the Q – quality factor of the system in the transient mode of metal capture by the rolls. The highest Q factor of the order of 130–170, which means the maximum amplitude and duration of the oscillations, is determined in stand No. 11. According to the results of wear measurements, it was found that in stand No. 11 there was maximum wear compared to the rest of the finishing stands, where the quality factor was about 20–50. In this case, in stand No. 11, the RMS vibration level in the idle mode was the highest. Therefore, the parameter Q can be considered a characteristic of the wear in the assembly units. The quality factor was estimated in the horizontal, vertical and axial directions; thereby comparing this parameter the wear of various components can be estimated.

In article [31], methods and devices for contactless (telemetric) measurement of torsional vibrations on rotating shafts of various machines and units were considered. Such systems can also implement the function of angular gaps diagnosing in the drivelines, since torsional moment in the driveline is highly informative about dynamic processes and can in turn be used to determine the transmission wear. For example, the method is known for diagnosing gaps by the time of a torsional shock pulse delay [32] measured in several points at the wide strip hot rolling mill. This method allows diagnosing of only the open parts of the gaps in the spindle joints. In work [33], various ways of diagnosing gaps in drivelines in transient operating modes are considered. However, the direct measurement of clearances allows obtaining reliable data on the total amount of wear in the joints. Therefore, special methods [34, 35] and systems are needed for such measurements.

There are known devices based on measuring the angular positions of transmission elements and tested on rolling mills for measuring wear (angular gaps) based on selsynes [36], optical [37] and potentiometric [38] angular position sensors of rotating shafts coupled through a detachable connection. To date, judging by the available data, devices and diagnostic methods of this kind are not widely used for diagnostics on industrial machines due to the inconvenience of installing and setting up the proposed sensors. Sensors, for example, selsyn (self-synchronized), should be installed at the ends of the shafts that is not practical in reality. The optical sensors of raster type require precise calibration and environmental protection from dust and humidity. Encoders must be also installed at the ends of the shafts that makes many elements of the drivelines inaccessible for diagnosing of angular gaps.

2 Methodology of Backlashes Measurement

The basis of the proposed method for diagnosing the full size of the angular gaps in the main lines of the rolling mill stands is the principle of sequential reversing of the direct current drive and rotation in both directions for 10–20 revolutions necessary to establish the stable rotation speed. In this case, for reliable closure of the gaps, a small load moment (up to 5–10% of the rolling moment) on the work rolls is created, for which they can be preloaded, e.g. by the pressure device of the rolling stand or applying pressure to the hydraulic cylinders of the balancing system or forced bending of the rolls. Sensors are mounted on drive shafts, for example, on both sides of a diagnosed gear coupling. Rotation of the driveline in the forward and reverse direction is produced by the electric motor; therefore, no additional devices are required. The basic principle of this method is discovered in [39]. In this paper, the issues of implementation is covered and results of industrial testing are represented.

2.1 Simulation of Measurement Method

The mathematical modeling of the proposed method is performed, bringing it as close as possible to real working conditions. For this, a diagram of the exemplary drivetrain was developed, containing the main elements and couplings with potential backlashes (see Fig. 4). The scheme consists of an electric motor rotor (eng), a pinion shaft (gw1) and a gear wheel (gw2) (z1 = 26, z2 = 188), a lower (pw1) and upper (pw2) rolls of a gear stand and a lower (wr1) and upper (wr2) work rolls. Angular clearances in the joints: motor coupling ($\Delta 1$), gearing of the reducer ($\Delta 2$), main coupling ($\Delta 3$), gearing of the gear stand ($\Delta 4$), lower ($\Delta 5$) and upper ($\Delta 6$) universal spindles.

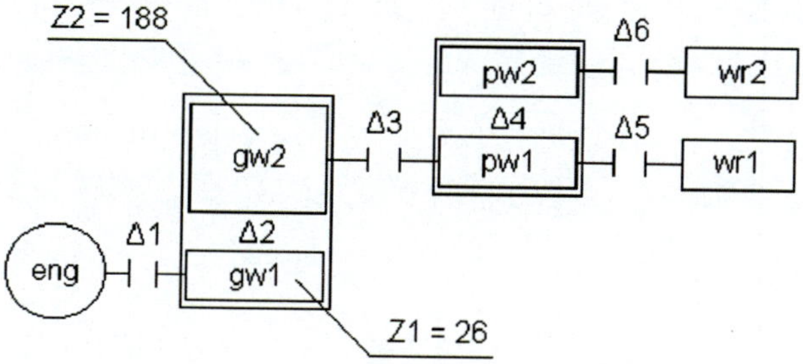

Fig. 4. The layout of the driveline and backlashes positions.

It was assumed that the driveline stopped at an arbitrary position. All angular clearances are closed in the main direction of rotation. All speed sensors were installed at the uppermost point of rotation of the main elements, the distances to the signal receivers and their response thresholds are the same. According to our scheme, we have 7 sensors, hence, the same number of measured signals. The values of all 6 gaps in the given diagram were arbitrarily set.

A hypothetical algorithm for the rotation of the motor rotor was simulated. First, its speed is set zero, then in a couple of seconds it accelerates to the set speed (40 rpm), rotates at this speed for 50 s, slows down, stands still for a few seconds, accelerates in the opposite direction, rotates at this speed for 57 s, slows down and stops (Fig. 5).

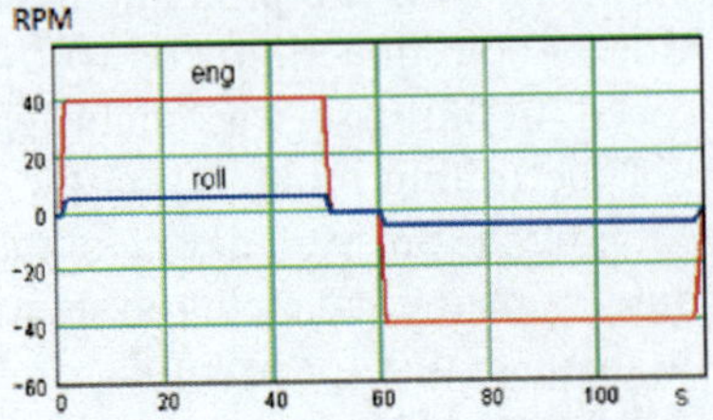

Fig. 5. Electrical motor and work roll speed diagram for one measurement test.

When all of the above conditions are met, we get such a picture of signals (Fig. 6). To determine the angular backlash in the joint, it is necessary to select the recording fragments in phase with each other, i.e. symmetric, where the axis of symmetry will be the moment of stopping. Such fragments are shown in Fig. 7 and Fig. 8. For the forward drive rotation in Fig. 7, the clearance in couplings are closed to one side of contact surfaces and the pulse offset is zero. The exception is coupling 2, where the speed is reduced and the signal offset (dT2a) inevitably occurs. The fragment at 65–67 s is in phase with this fragment. In this case, the backlashes in all joints are completely open and we see a shift in impulses along the entire driveline (Fig. 8).

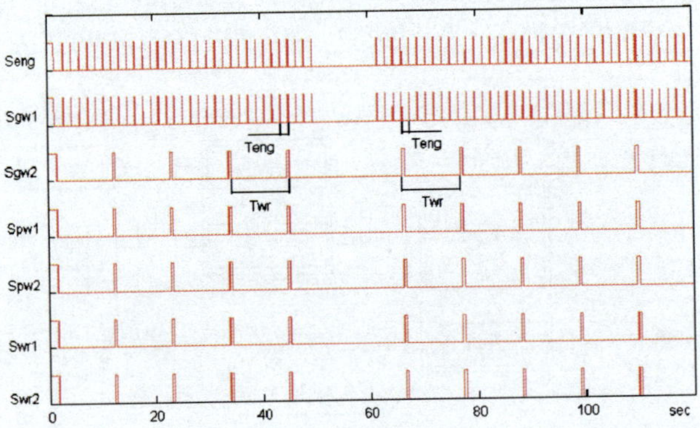

Fig. 6. The simulated signals of rotation speed sensors of the driveline.

To calculate the backlash in all joints, we will use such dependences where the result will be in radians:

$$\Delta_1 = \frac{dT1r}{Teng};\ \Delta_2 = \frac{dT1r - dT2a}{Teng};\ \Delta_3 = \frac{dT3r}{Twr};$$
$$\Delta_4 = \frac{dT4r}{Twr};\ \Delta_5 = \frac{dT5r}{Twr};\ \Delta_6 = \frac{dT6r}{Twr} \tag{2}$$

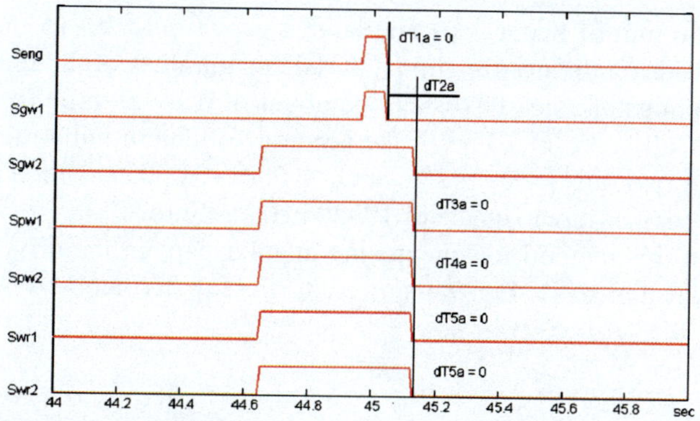

Fig. 7. Fragment of signals measurements at the forward drive rotation.

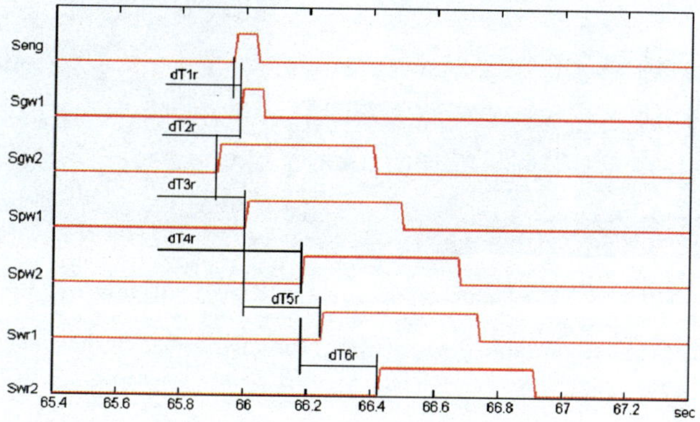

Fig. 8. Fragment of signals measurements at the backward drive rotation.

Thus, we have tested on a mathematical model a method for measurement of angular clearances, which is implemented in practice for the drivelines wear diagnostics of industrial machines mainly rolling mills.

3 Digital Device for the Backlashes Measurement

The proposed method for the diagnosis of gaps is based on the use of two standard speed sensors installed on the side surfaces of the shafts. Their signals are recorded and show the time between successive pulses from the different points in the driveline. By comparing the difference in time intervals between pulses from the speed sensors during rotation in the forward and reverse drive directions, the total value of the gaps is determined, and the open value of the angular gaps affecting the dynamics of the driveline is determined from the values of the time intervals during idling and under load.

With a minimum sampling rate of device 0.01 s (100 Hz), and rotation speed of 1 rev/s, the minimum detectable angular clearance is $0.01 \times 2\pi = 0.0628$ rad (3.6 grad), which even with a relatively small shaft of 400 mm diameter is about $\Delta = 200$

× 0.0628 = 12.56 mm of linear gap. The largest gaps (up to 10–15 mm) are formed in the spindles heads on the bronze slip pads. The magnitudes of the linear gaps in the gears of the couplings and gearboxes can be about of 0.2–1.0 mm. To diagnose such gaps at a motor shaft speed of 10 rev/s, the accuracy of determining the time intervals between pulses of two sensors should be about 0.0001–0.0005 s, which corresponds to a sampling frequency of input signals of 10–20 kHz or more.

To implement this method for diagnosing angular gaps in the drivelines, a digital device has been developed (see Fig. 9) using contactless shaft rotation sensors as primary transducers.

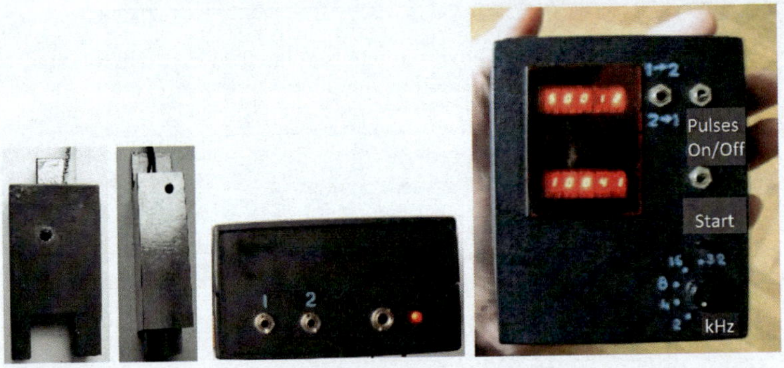

(a) (b) (c)

Fig. 9. Sensors (a) and portable device (b), (c) for angular backlashes measurements.

The portable digital device includes:

- two optical sensors or Hall sensors (see Fig. 9a) with cables;
- two optical tags or two small permanent magnets mounted on the shaft;
- device itself with a cable for communication with a computer (see Fig. 9b,c).

Sensors are connected to the device via cables, with male connectors (1–2) on the side panel (see Fig. 9b). There is also a connector for output signal to a computer and recording digital signals from sensors. Signal output is indicated by LED.

The following elements are located on the front panel of the device (Fig. 9c):

- 5-digit indicator (upper) difference in the number of pulses of sensors;
- 5-digit indicator (lower) of the number of pulses per shaft rotation;
- master channel switch ("1 → 2", "2 → 1") to control the pulse generator;
- clock output switch ("TI") to a computer;
- instrument power switch ("On");
- clock count start button ("Start");
- clock frequency switch ("2… 32 kHz").

The functional scheme of the device is given in Fig. 10 and it works as follows. The signals from the speed sensors come into the device, where after the driver and the channel switch are fed to the inputs of the clock generator. The frequency of clock pulses is selected in advance in the range of 2…32 kHz so that when the speed of rotation of the drive is set, there is no overflow of the 5-digit pulse counter per shaft revolution. The generator clock switch also sets the sensitivity of the device depending on the size of the gaps. Use the "Start" button to count pulses when the shaft rotates in the forward and reverse direction. Switch "TI" is used to output clock pulses as a test for checking the line of communication with the computer.

Angular gaps are determined by calculating the ratio of the value on the upper indicator to the value on the lower indicator. Linear clearance is determined by the initial diameter of the engagement circle of the coupling or gearbox.

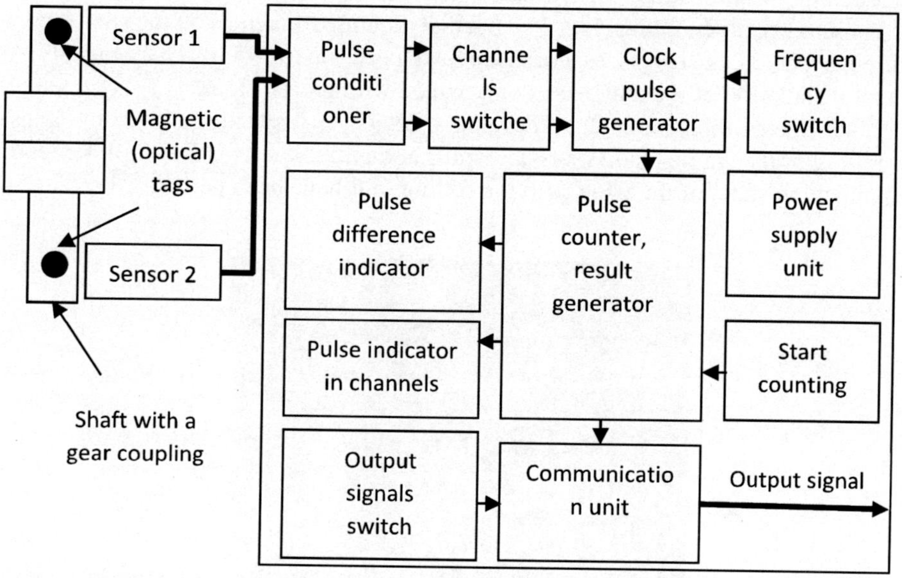

Fig. 10. Functional scheme of the device for measuring angular gaps in the drivelines.

The device can work as a separate device on battery power, and as part of a measuring and computing complex based on a computer. To input signals from the instrument and write to the hard disk of the computer, use the cable and connector on the back panel of the instrument. In this case, signals can be input even without a special ADC board, and through the line-in input of a computer sound card. At the same time, in connection with the well-known property of sound cards to differentiate input signals, instead of each rectangular pulse, two pulses of different polarity are obtained, a positive pulse at the leading edge and a negative pulse at the falling edge of the original signal, which is then taken into account when processing in software. The ability to input signals from the device through a sound card greatly simplifies and reduces the cost of system configuration, since the cost of ADC converters is quite high and can be comparable to the cost of the computer itself.

At this stage of development, the portable diagnostic device is realized on the non-programmable logical elements. Further implementation is supposed to be based on a microcontroller with an LCD display, which will allow more comfortable visualization and calculating gaps directly in the device from an assigned by user diameter value, as well as reducing the size and power consumption of the device by several times.

4 Industrial Trials

The developed device for backlashes metering is also tested on several drivelines of industrial rolling mills, as well as on auxiliary machines (strip coiler), where it showed good performance. Shaft rotation sensors are installed on the equipment using conventional magnetic tripod mounts at a distance of 10–20 mm from the shaft, on which two magnetic tags were installed with a diameter of 10–15 mm on both sides of the diagnosed unit, e.g. gear coupling (Fig. 11). Special mounting magnets on the shafts are not required, because cage drives, as a rule, are low-speed (up to 3–5 rot/s). The shaft in the place of installation of the magnets is only wiped with grease cloth.

With the help of this method and developed device, an inspection of the drivelines of cold rolling mills (single-stand reversible mills and multi-stand tandem mill, tempering mill), finishing stand of the wide strip hot rolling mill have been performed.

Fig. 11. Measurement of gaps in the gear coupling.

The small dimensions of the tags on the shafts allow them to be installed on the shortest sections (up to 10 mm), for example, between the coupling sleeve and the gearbox housing, on the roll necks and spindle head. The results of measurements and data processing are shown in Fig. 12.

Measurements of time delays between impulses during reverse rotations showed the exact compatibility with the values of gaps measured with manually after gear coupling disassembling for repair. The estimated accuracy of gaps measurement by the tested method is about 0.1–0.2 mm depending on conditions and machine design. The main factor affecting accuracy is the standard deviation of rotation speed, which is easier to reduce in case of idle mode of machine rotation.

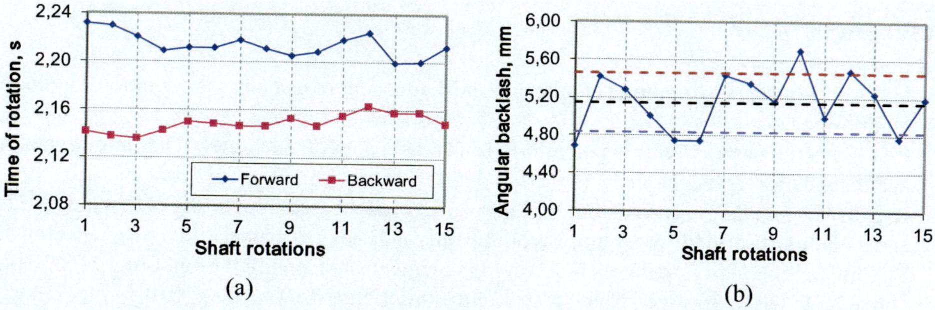

Fig. 12. Time diagrams of backlashes measurement in the driveline: (a) time of rotations; (b) calculated angular backlash (mean value 5.15 mm, standard deviation ±0.31 mm).

5 Discussion and Conclusions

A review is represented of existing methods and devices for measuring gaps in transmissions of mobile vehicles and stationary machines, including drivelines for rolling mills. Based on the analysis of existing methods and devices the contactless measurement of angular gaps on rotating shafts is proposed and tested in laboratory tests and industrial trials.

The designed device and method of backlash measurement can also be implemented in monitoring systems as a tool of torsional loads in the machines. An important function of the proposed method and device is the ability to evaluate (continuous adaptation) of the stiffness (compliance) of the driveline sections. Since the inertial parameters of the equipment practically do not change during operation, the obtained stiffness parameters allow on-line identification of the dynamic model and perform peak torques calculations in those elements where strain gauges and torque meters are not possible to install. This model-based approach provides more possibilities for accurate calculations of cyclic fatigue and remaining useful life prediction. In addition, the measured value of torsional stiffness itself is a diagnostic parameter of the technical condition of the driveline elements.

In general, angular and radial (linear) gaps greatly affect the dynamic loads and cause the sudden failures in the drivelines, hence, they should be considered as the most critical parameters in maintenance of the rotating equipment working under non-stationary loading conditions, especially in reversal regimes with impulsive and step-like loads application. It is necessary to diagnose both the full gap of couplings in idle mode and opened part of backlashes. A digital instrument has been developed for measuring gaps in the drivelines with reversible drive. To diagnose the open part of the full gap, it is advisable to use transient modes of operation of the machines (acceleration-deceleration, capture-release of material by working tool).

The laboratory testing and industrial trials proved the ability to measure angular backlashes with an accuracy up to 0.1 mm. The results are not dependable on rotational speed deviations and applicable for wide range of speeds.

References

1. ISO 13373 (all parts) Condition monitoring and diagnostics of machines—Vibration condition monitoring
2. ISO 13379 (all parts), Condition monitoring and diagnostics of machines—Data interpretation and diagnostics techniques
3. Krot, P., Prykhodko, I., Raznosilin, V., Zimroz, R.: Model based monitoring of dynamic loads and remaining useful life prediction in rolling mills and heavy machinery. In: Ball, A., Gelman, L., Rao, B.K.N. (eds.) Advances in Asset Management and Condition Monitoring. SIST, vol. 166, pp. 399–416. Springer, Cham (2020). https://doi.org/10.1007/978-3-030-57745-2_34
4. Krot, P.V.: Dynamics and diagnostics of the rolling mills drivelines with non-smooth stiffness characteristics. In: Proceedings of the 5th International Conference on Nonlinear Dynamics ND-KhPI2016, Kharkov, Ukraine, pp. 21–24 (2016)
5. Krot P.V., Putnoki A.Y., Klevtsov O.M., Ermolenko A.A.: Experimental studies and industrial testing of methods for reducing dynamic loads in the gearbox drivelines of roughing stands NTLS 1680. In: Proceedings of the V Congress of Rolling, 21–23 October 2003, Cherepovets, pp. 523–529 (2004)
6. Belyansky, A.D., Voskanyants, A.A., Gartsman, S.D., et al.: PatentSU 910247. IPC B21B 35/14. The way to configure the spindle device. NKMZ, No. 2956371/22-02, Appl. 07.21.1980, Publ. 03.07.1982. Bul. №9
7. Krot, P., Zimroz, R., Sliwinski, P., Gomolla, N.: Safe operation of underground mining vehicles based on cyclic fatigue monitoring of powertrains. In: Lesiuk, G., Szata, M., Blazejewski, W., Jesus, A.M.d., Correia, J.A. (eds.) Structural Integrity and Fatigue Failure Analysis. VCMF 2020. Structural Integrity, vol. 25, pp. 283–292. Springer, Cham (2022). https://doi.org/10.1007/978-3-030-91847-7_26
8. Krot, P.V.: Dynamical processes in a multi-motor gear drive of heavy slabbing mill. J. Vibroeng. **21**(8), 2064–2081 (2019)
9. Krot, P., Korennoi, V., Zimroz, R.: Vibration-based diagnostics of radial clearances and bolts loosening in the bearing supports of the heavy-duty gearboxes. Sensors **20**(24), 7284 (2020)
10. Krot, P.V.: Transient torsional vibrations control in the geared drive trains of the hot rolling mills. In: Proceedings of the IEEE Multi-conference on Systems and Control, St. Petersburg, Russia, pp. 1368–1373 (2009)
11. Muszynski, R., Deskur, J.: Damping of torsional vibrations in high-dynamic industrial drives. IEEE Trans. Industr. Electron. **57**(2), 544–552 (2010)
12. Gasiyarov, V.R., Khramshin, V.R., Voronin, S.S., Lisovskaya, T.A., Gasiyarova, O.A.: Dynamic torque limitation principle in the main line of a mill stand: explanation and rationale for use. Machines **7**(4), 76 (2019)
13. Wang, K.-S., Li, Z., Braaten, J., Yu, Q.: Interpretation and compensation of backlash error data in machine centers for intelligent predictive maintenance using ANNs. Adv. Manuf. **3**(2), 97–104 (2015). https://doi.org/10.1007/s40436-015-0107-4
14. Radionov, A.A., Gasiyarova, O.A., Baskov, S.N.: Angular clearance monitoring for mechanical transmissions of rolling mill stands. In: Radionov, A.A., Gasiyarov, V.R. (eds.) Proceedings of the 7th International Conference on Industrial Engineering (ICIE 2021). LNME, pp. 679–689. Springer, Cham (2022). https://doi.org/10.1007/978-3-030-85230-6_80
15. Gasiyarova, O.A., Karandaev, A.S., Erdakov, I.N., Loginov, B.M., Khramshin, V.R.: Developing digital observer of angular gaps in rolling stand mechatronic system. Machines **10**, 141 (2022)
16. Li, M., Liang, R., Zhang, Y., Peng, C., Mu, D., Wan, Z.: On-line measurement method of transmission backlash based on angular velocity and double-end angular position information. Meas. Control **54**(1–2), 65–72 (2021)

17. Sander-Tavallaey, S., Saarinen, K.: Backlash identification in transmission unit. In: IEEE Control Applications, (CCA) & Intelligent Control, (ISIC 2009), pp. 1325–1331 (2009)
18. Sarkara, N., Ellis, R.E., Moore, T.N.: Backlash detection in geared mechanisms: modeling, simulation, and experimentation. J. Mech. Syst. Signal Process. **11**(3), 391–408 (1997)
19. Guzinski, J., Diguet, M., Krzeminski, Z., Lewicki, A., Abu-Rub, H.: Application of speed and load torque observers in high-speed train drive for diagnostic purposes. IEEE Trans. Industr. Electron. **56**(1), 248–256 (2009)
20. Li, X., Zhao, D., Xie, F., Wu, S., Li, X.: Experimental investigations of the dynamic responses of a multi-link mechanism with revolute clearance joints. Adv. Mech. Eng. **13**(4), 168781402110125 (2021)
21. Gomez, J.L., Khelf, I., Bourdon, A., André, H., Rémond, D.: Angular modeling of a rotating machine in non-stationary conditions: application to monitoring bearing defects of wind turbines with instantaneous angular speed. Mech. Mach. Theory **136**, 27–51 (2019)
22. Tian, Q., Flores, P., Lankarani, H.M.: A comprehensive survey of the analytical, numerical and experimental methodologies for dynamics of multibody mechanical systems with clearance or imperfect joints. Mech. Mach. Theory **122**, 1–57 (2018)
23. Chongyi, G., Jianxiong, L., Guojun, D.: Comparative research of nonlinear torsional vibration in the drive system with clearances. In: 54th Annual Conference of the Society of Instrument and Control Engineers of Japan (SICE), pp. 1003–1006 (2015)
24. Gaponov, V.S., Muzykin, Y.D., Tatkov, V.V., Putnoki, A.Y., Voytovich, A.I., Fedorenko, I.M.: Measuring complex for determination of wear of the lateral surface of the teeth. Visnyk NTU "KhPI". Ser. New Solutions Mod. Technol. **43**(1016), 51–56 (2013)
25. Atamanov, V., Kudrjavtsev, E., Pronyakin, V., Guljaev, A.: On the determination of the gear teeth wear using an inductive sensor. Sci. Educ. Bauman MSTU **6**, 10–21 (2015). https://doi.org/10.7463/0615.0779329
26. Leduc, H., Bozzonetti, C.: FR 2958028 A1. IPC G01B5/166 Gear backlash measuring method for automobile transmission, Appl. 2010-03-26, Publ. 2012-05-25
27. Podkopaev, V.V., Podkopaev, S.V.: Patent SU 1422055 A1. IPC G01M 13/02. The method of determining the wear of the main transmission line of vehicles, No. 4218738/30-11, Appl. 03.30.1987, Publ. 09.07.1988. Bul. No. 33
28. Chistyakov, Y.P., et al.: Patent RU 2147474 C1. IPC B21B 38/00. A device for determining wear in friction pairs of a stand pressure device. No. 98112838/02, appl. 07.06.1998, publ. 04.20.2000. Bul. №4
29. Gromakovsky, D.G., Averkieva, V.I., Yafyasov, R.A., et al.: Method of diagnosing friction units. Patent SU 1441255 A1. IPC G01N 03/56. No. 429898/25-28, Appl. 13.04.1987, Publ. 11.30.1988. Bul. №44
30. Fedorov, P.F., Nosov, V.L., et al.: Determination of the technical condition of the mill stands 2500 of OJSC MMK according to the results of the study of the quality factor of oscillations of the roll system. Prod. Rolled Prod. **5**, 20–22 (2002)
31. Krot, P.V.: Telemetry monitoring systems of dynamic loads in the drive lines of rolling mills. Vibr. Mach. Meas. Reduction Prot. **1**, 30–41 (2008)
32. Verenev, V.V., Yunakov, O.M., Dalichuk, A.P.: Method of determining technical condition of driveline in rolling stand. Patent UA 70137A. IPC G01M 7/00. No. 20031212642, Appl. 12.26.2003, Publ. 15.09.2004. Bul. № 19
33. Verenev, V.V., Bolshakov, V.I., Putnoki, A.Y., et al.: Diagnostics and Dynamics of Rolling Mills, p. 144. IMA-Press, Dnepr (2007)
34. Bolshakov, V.I., Krot, P.V., Korennoi, V.V., et al.: A method for diagnosing bearing supports of the main driveline of a rolling mill. Patent UA 79681, IPC G01M 13/00. No. 20050008869, Appl. 19.09.2005, Publ. 07.10.2007. Bul. №10

35. Bolshakov, V.I., Krot, P.V., Korennoi, V.V., et al.: A method for diagnosing wear of transmission elements of rolling stands. Patent UA 79859. IPC B21B 28/00. No. 20050008854, Appl. 19.09.2005, Publ. 07.25.2007. Bul. №11
36. Kotsar, S.A., Polyakov, B.A., Belyansky, A.D., et al.: Reducing angular clearances and dynamic loads when metal is gripped by rollers. Bull. Chermetinformation, Ferrous Metall. **2**, 42 (1979)
37. Plakhtin, V.D., Danilov, L.I., Moskvitin, S.A.: A device for diagnosing the state of the connection of the rolling mill shafts. Bull. Chermetinformation Ferrous Metall. **20**, 45 (1982)
38. Podkovyrin, E.Y., Skichko, P.Y.: Sensor for measuring gaps in the joints of rotating parts. Collect. Sci. Pap. "Dyn. Metall. Mach. **31**, 161–165 (1969)
39. Krot, P.V.: Methods and equipment for measuring wear in rolling mill drivelines. Metall. Process. Equip. **2**(12), 45–53 (2008)

Author Index

A
Augustyn, Damian 194, 203

C
Caso, Enrique 52

D
De-Juan-De-Luna, Ana 52
Díez-Ibarbia, Alberto 52
Dobaj, Krzysztof 85
Dworakowski, Ziemowit 13, 75, 123

F
Fernández-del-Rincón, Alfonso 52
Fidali, Marek 95, 194, 203

G
Gajdzik, Marcin 144
García-Fernández, Pablo 52
Gnacy-Gajdzik, Anna 144
Grządziela, Andrzej 21

H
Hetmanczyk, Mariusz 176

J
Jagoda, Jerzy 176

K
Kania, Konrad 166
Kluczyk, Marcin 1, 21
Kolar, Davor 1
Korbiel, Tomasz 115
Korennoi, Volodymyr 212
Kot, Adam 32
Król, Roman 63
Krot, Pavlo 212

L
Łatas, Waldemar 32
Lisjak, Dragutin 1
Lorencki, Jakub 105

M
Machynia, Adam 13, 75
Muślewski, Łukasz 1

P
Pająk, Michał 1
Pawlik, Paweł 166
Pedot, Massimiliano 131
Peruń, Grzegorz 42
Przystałka, Piotr 144
Przysucha, Bartosz 166

R
Radkowski, Stanisław 105
Roczek, Krzysztof 95
Roemer, Jakub 13, 75
Rogala-Rojek, Joanna 176

S
Sánchez-Espiga, Javier 52
Sobczak, Michał 13, 75
Stankiewicz, Krzysztof 176
Sternal, Kamil 186
Szrek, Jarosław 212

T
Timofiejczuk, Anna 131, 144

V
Vidan, Pero 21

Z
Zimroz, Radosław 212
Żymełka, Szymon 154

A. Puchalski et al. (Eds.): ICTD 2022, ACM 21, p. 229, 2023.
https://doi.org/10.1007/978-3-031-31719-4